# LIFE ITSELF ●

EXPLORING THE REALM OF THE LIVING CELL

LIFE
ITSELF

●

BOYCE RENSBERGER

ILLUSTRATIONS BY NIGEL ORME

OXFORD UNIVERSITY PRESS
NEW YORK   OXFORD

Oxford University Press

Oxford    New York
Athens   Auckland   Bangkok   Bogotá   Buenos Aires   Calcutta
Cape Town   Chennai   Dar es Salaam   Delhi   Florence   Hong Kong   Istanbul
Karachi   Kuala Lumpur   Madrid   Melbourne   Mexico City   Mumbai
Nairobi   Paris   São Paulo   Singapore   Taipei   Tokyo   Toronto   Warsaw

*and associated companies in*

Berlin   Ibadan

Copyright © 1996 by Oxford University Press

First published by Oxford University Press, Inc., 1996

First issued as an Oxford University Press paperback, 1998

Oxford is a registered trademark of Oxford University Press

Library of Congress Cataloging-in-Publication Data
Rensberger, Boyce.
Life itself: Exploring the realm of the living cell / by Boyce Rensberger.
p. cm.   Includes bibliographical references and index.
ISBN 0-19-510874-4
ISBN 0-19-512500-2 (Pbk.)
1.Cytology—Popular works.   I. Title.
QH581.2.R46   1996
574.87—dc20   96-33679

10 9 8 7 6 5 4 3 2 1

Printed in the United States of America

For Linda

# PREFACE AND
# ACKNOWLEDGMENTS ●

So many people helped me to do this book that in the course of research-
ing and writing it, I have made a most gratifying network of friends in
cell and molecular biology.

I owe a special debt to Robert D. Goldman, for it was he who introduced me to the mysteries of cells. Through the generosity of the Science
Writers Fellowship Program at the Marine Biological Laboratory in
Woods Hole, Massachusetts, I was allowed to spend the summer of 1987
immersed in the fascinating realm of the cell. Goldman was directing
MBL's famed "Physiology" course that summer—an intensive eight
weeks of total immersion in the world of cells and in state-of-the-art
methods of studying them. The course is intended for graduate students
and Ph.D.s, but MBL gives fellowships to a few lucky science writers to
attend as well. I did immunofluoresence microscopy, poured and ran gels,
made cDNA libraries, and carried out many other lab methods I had
barely heard of before.

On the first day of the course, Goldman put some live, baby hamster
kidney cells under a microscope, aimed a video camera through the
eyepiece, and showed the class a close-up view of a single cell. To me it
seemed not some disembodied fragment of a mammal but an organism, a
microscopic creature in its own right. Within the nearly transparent cell, I
could easily see tiny objects moving around, shuttling here and there as
the cell performed the myraid interactions that, taken together, are life.
Life itself, I realized, was not a blob of homogeneous protoplasm—"the
stuff of life," as I remember some old textbook putting it—but a bustling
beehive. I was hooked.

Many times during the course of writing this book, Goldman helped
with ideas and concepts, names and numbers, facts and figures, encour-
agement and confidence. When I wanted pictures of real cells, he taught

me to use the microscopes to make many of the pictures myself. For one privileged week I sat at light microscopes and electron microscopes in Goldman's Chicago laboratory as his staff members Satya Khoun and Michelle Montag-Lowey brought me slide after slide of cells to examine and photograph. Manette McReynolds helped operate the scanning electron microscope, which gives such lifelike three-dimensional views.

At the end of the week, I had more than 600 negatives of cells. Goldman examined them over a lightbox, throwing out one after another as out-of-focus or badly exposed. Ruthlessly, he tossed away images I had carefully obtained of interesting looking objects that turned out to be the laboratory equivalent of fingers in front of the lens. In the end, however, a handful of pictures survived, and a few of them are in this book

My experience at Northwestern was made possible by a generous grant from the Alfred P. Sloan Foundation, which has long been an advocate and supporter of science writing. Without the foundation's financial support it would have been impossible to carry work on this book to completion. I am deeply grateful.

Too, I must express my debt to Robert G. Kaiser and Leonard Downie, my editors at the *Washington Post,* who gave me the time to go to Woods Hole and who allowed me to write a long, five-part series on cell biology for the *Post*—a series that did not meet the conventional definition of news but that simply told readers about the inner workings of cells. Len and Bob put it on the front page of the paper every day—the only time, I suspect, that such things as immunofluorescence images of cytoskeletons have ever made "Page One" of an American daily newspaper. That series elicited a huge outpouring of response from fascinated readers, and this book is a direct outgrowth of that series.

At Garland Publishing, I am indebted to Elizabeth Borden, who saw merit in this book after many other publishers turned it down, and to Ruth Adams, who managed the production with grace and understanding. I thank Nigel Orme for the illustrations that add so much meaning to the text and visual appeal to the pages. And I thank Bruce Alberts for bringing my manuscript to Garland's attention.

I must mention Bob Goldman again, for I owe him another kind of debt. He decided which scientist applicants to accept into the Physiology course the summer I was there, and one of them was Linda A. Thomas, a Ph.D. candidate at the University of California, San Francisco. Her enthusiasm for science and cell biology was irresistible, as was everything else about her. We are now happily married.

In the course of doing this book, a great many experts have been most generous with their time. One of the great joys of being a science writer is being able to command tutorials from the top experts in any field of research. I do not remember a scientist ever refusing to explain something to me or ever failing to keep reexplaining it until I understood—or, at least, thought I understood.

Among those who gave me their time and knowledge for this book I remember, and thank, these: Bruce Alberts, Guenter Albrecht-Buehler, Mina Bissell, Gary Borisey, Dennis Bray, B. R. Brinkley, Kay Broschat, Keith Burridge, Rex Chisholm, Shiela Counce, Caroline Damsky, George Dessev, Harold Erickson, Susan Fisher, Yoshio Fukui, Ann Goldman, Robert D. Goldman, Allan L. Goldstein, James Hagstrom, Rob Hay, Millie Hughes-Fulford, Tim Hunt, Shinya Inoue, Alexander D. "Sandy" Johnson, Marc Kirschner, Hynda Kleinman, Edward Korn, Julian Lewis, David McClay, Tim Mitchison, Bruce Niklas, Robert Palazzo, Tom Pollard, Martin Raff, Michael Sheetz, Ray Stephens, Richard Tasca, Linda Thomas, Lou Tilney, John Tooze, Richard Vallee, Karen Vikstrom, Byron Waksman, and Peter Walter.

Several people went even further in trying to help me make the information in this book as accurate as possible. They read all or part of the manuscript and offered corrections and suggestions. I'd like to name and thank them: Bruce Alberts, Guenter Albrecht-Buehler, Dennis Bray, Rex Chisholm, Robert D. Goldman, Allan L. Goldstein, Rob Hay, Alexander D. "Sandy" Johnson, Julian Lewis, Martin Raff, Michael Sheetz, Richard Tasca, Linda Thomas, Ray Stephens, John Tooze, Byron Waksman, Peter Walter, and Stuart Yuspa.

Any mistakes that remain are, of course, my fault. To all cell and molecular biologists I must make a humbling acknowledgment: You're faster than I am, and you've got me outnumbered. As you know better than I, this is a fast-moving field. In the newspaper business we'd call it a fast-breaking story. There are bound to be instances in which I have said something was unknown, and now you know. Or, worse yet, where the explanation I gave is no longer the best one, given new data. And, of course, the research is extraordinarily complex.

Between the time I started this book, in late 1989, and when I last let go of the manuscript—early 1995—I was frequently coming across new findings and having to update. Or discovering things that I had misunderstood. In one sense, I would like to hope that nothing new has been learned since I finished writing and that readers could be assured this book was up-to-the-minute. But in reality I know that couldn't be the case and, actually, I'm glad it's not. There are still so many mysteries left to be solved in our understanding of how life works, and I, for one, am dying to know the answers.

Our civilization does not support basic science solely that scientists themselves may satisfy their own curiosities. We taxpayers support it so that all may share in the knowledge. To be sure, we pay for biomedial research with some confidence that it will eventually yield practical benefits. That is the primary goal of the National Institutes of Health, the extraordinary federal agency that funded the discovery of so much of what is in this book, either through its own laboratories or by giving grants to cell biologists and molecular biologists at other research centers around the country and elsewhere.

But that practical aim should not blind us nonscientists to the fact that the knowledge itself is well worth having, quite apart from its practical use. Human beings are by nature curious and questioning. The scientists who get the grants to do the research are acting as the agents of our curiosity. When they make a discovery, we want to shout "Eureka!" with them. The details that they discover and that we subsequently learn, enrich and ennoble our lives as surely as do the symphonies and poems and paintings of the greatest artists. I hope that this book can help to extend that intangible "spinoff" of cell and molecular biology to more people.

*Bethesda, Md.*                                                                 B. R.
*November 1995*

# CONTENTS ●

# LIFE ITSELF ●

Clouds of cold fog billow up through a circular hatch in the top of a stainless steel tank as Rob Hay pulls out the lid and its one-foot thickness of styrofoam insulation. As the fog rolls down to the floor, Hay peers into the dark tank, where the temperature is always 321 degrees below zero Fahrenheit.

It is kept so cold because that is a temperature at which life, normally warm and pulsing with activity, abandons its vital dance and enters limbo—but without dying.

Inside Rob Hay's stainless steel tank are about 60 billion life-forms gathered from all over the planet. They wait in chilly repose. They neither eat nor sleep. They do not breathe. Not even the simplest chemical reactions of metabolism take place inside their bodies. By any conventional definition of life, the creatures have surrendered their claim on it.

And yet by those definitions, anyone can work a miracle: Simply reach into the tank and take out one of the 30,000 hermetically sealed glass vials, or ampoules, each about an inch long and each holding about two million of the inanimate creatures in less than a teaspoonful of frozen fluid. Now let it warm to body temperature. From their icy limbo, the tiny creatures will undergo resurrection in minutes. They will, in the language of an age that could not have imagined what is routine in almost every biomedical laboratory today, "come back to life." Some of the little organisms begin moving around, crawling over the inside surface of their container. They will begin feeding on the nutrients dissolved in the now-melted liquid that surrounds them. And—the surest proof that they are truly alive—the creatures will begin reproducing.

The organisms in the ampoules are cells—the fundamental units of life, the microscopic building blocks of which all living organisms are made. They have been removed from the bodies of thousands of different animals of nearly a hundred species. In some of the frozen ampoules are kidney cells from a mouse. In others are skin cells from a chimpanzee. There are turkey blood cells, armadillo spleen cells, iguana heart cells—dozens of different kinds of cells from scores of different species. And, of course, there are cells of human beings. More than 20,000 frozen ampoules hold about 40 billion human cells in suspended animation.

Inside ampoules marked "ATCC CCL 72," for example, are skin cells taken in 1962 from a nine-month-old baby girl. The child has long since died of a birth defect, but her cells survive, ready to "come back to life" any time anybody removes an ampoule and warms it up. In "ATCC HTB 138" are brain cells removed from a seventy-six-year-old man in 1976. He too is now dead, but not his brain cells. In ampoule "ATCC CCL 204" are lung cells from a thirty-five-year-old man who died in 1979 and whose body was maintained so that its organs could be transplanted. Before the life support system was switched off, researchers bestowed a kind of immortality on a tiny piece of his lung. They sliced out a chunk containing a few thousand lucky cells and granted them the opportunity to live on, growing and reproducing themselves independently for decades.

The most famous cultured cells, or cell line, in the annals of biomedical research are those in "ATCC CCL 2," known worldwide as HeLa cells because they came from Henrietta Lacks, a thirty-one-year-old Baltimore woman who died in 1951 of cervical cancer. The cells from her tumor have been multiplying in laboratories all over the world ever since. At one time HeLa cells were among the most popular with researchers, who, like gardeners sharing cuttings from a favored plant, freely passed a few cells from their own colony to other researchers. So many dishes and flasks of the HeLa cell line are now alive around the world that it is estimated they weigh far more than Henrietta Lacks ever did.

Some 1,400 other cell lines, representing more than 70 different types of tissue from hundreds of human beings, are also held frozen in sealed glass ampoules inside stainless steel tanks like the one Rob Hay is looking into. Hay is a biomedical scientist and director of cell biology at one of the most unusual organizations in the world, the only one of its kind in the United States. It is a bank of living, though mostly frozen, cells called the American Type Culture Collection, which occupies a nondescript, red brick building in Rockville, Maryland, a suburb of Washington, D.C. The ATCC, as it is known throughout the biomedical research community, every year ships more than 50,000 ampoules of frozen cells to scientists around the world. The ampoules are packed in dry ice chunks inside a thick styrofoam cooler.

"When they first started shipping cells years ago, it was considered a pretty weighty decision," recalls Robert E. Stevenson, a longtime ATCC director. "After all, they wondered, since it was human cells, would anybody be afraid that human life was being sent through the mail? They even checked with the Catholic Church. It turned out nobody really had a problem with it once they understood what was going on."

When researchers receive their cells, they perform routine resurrections, thawing the cells, transferring them to new bottles with fresh nutrient-rich broth, and rearing successive generations of multiplying cells.

On the cutting edge of modern medical research, cultured cells are the most intimately studied life-forms. Biologists still use mice, white rats, guinea pigs, and fruit flies. But more and more it has become evident that many of the most challenging questions—from the practical desire to conquer disease to the purely intellectual quest to understand how life works—can only be answered with a detailed knowledge of the innermost workings of individual cells.

Human life, biologists now know, is really the sum of the lives of many individual cells organized in specific ways. And all disease is the result of processes that go awry within cells or among cells. Cancer and the common cold, heart disease and hay fever, even AIDS and arthritis, all afflict people because of failed mechanisms within cells. This is why biologists say basic research—aimed at understanding how cells work rather than at conquering any specific disease—promises knowledge that can be used to attack many, and perhaps all, diseases.

Sickle-cell anemia, for example, is the result of the slightly flawed shape of one kind of molecule in a person's blood cells. The errant molecule warps the red blood cell's normal doughnut shape into a crescent, or sickle, shape. Some forms of diabetes are caused by faulty molecules called receptors that are embedded in the membranes that enclose cells. These molecules are specialized gatekeepers, intended to recognize only insulin molecules and, when one comes along and binds to it, they will relay the appropriate signal to the cell's interior. With faulty shapes, the molecules fail to recognize insulin and the body becomes as sick as if it had no insulin at all.

As scientists probe the innermost workings of these microscopic components of the human body, they are beginning to understand one of the most profound mysteries ever contemplated—the nature of life and how it works.

Until the mid-nineteenth century, scientists were ignorant of the existence of cells. Life seemed to come only in units of the whole organism—a cat, a bird, a human being. Many held that the fundamental components of organisms were structural members—fibers and vessels, specifically; these strings and tubes, they believed, somehow grew like crystals, enlarging and changing shape until they attained adult proportions. Skin was seen as a modified tube, closed at the ends. Muscle was obviously bundles of fibers. The various internal organs were simply vessels of different shapes, no more amazing than a pipe or a pot.

The orderly predictableness with which the body grew from conception—always forming a variety of specialized organs structured in the same way—was attributed to a theoretical property called the "vital force." It was often thought of as the breath of life that God bestowed on Adam after fashioning his body from clay. Philosophers, arguing on behalf of a theory called "vitalism," said that scientists could probe only so far in seeking to understand how life works. Then they would run up against something that was inherently unknowable, something beyond the realm of science. The presence of this entity in a human body was said to be that of the soul—a phenomenon that mortals could not hope to study nor understand. In the early eighteenth century George E. Stahl, the first of a long line of German scholars to dominate early cell biology, championed an explicit version of vitalism called "animism." He taught that all the physical parts of the body were passive and that motion, or animation, came from the soul. Although it is tempting nowadays to ridicule such views, in the days before there was any significant knowledge about cells, the theory of vitalism was mainstream science. In that age when modern science was just beginning, scientists were not at all uncomfortable mixing naturalistic explanations with supernatural ones.

Death in those days was thought of as the departure of the vital force, the exiting of the soul, the giving up of the ghost, or spirit. All the structural forms of the body might seem to remain intact in the corpse, but once the vital force had gone, the body was no longer animated but now inert and subject to decay and corruption.

Over the ensuing centuries, however, biologists would probe deeper into the living things they studied and gradually find that most of the phenomena they attributed to mystical forces could, in fact, be explained by entirely natural processes. As scientists gradually demanded more and more evidence for their theories, the concept of vitalism slowly faded. In 1838 and 1839, for example, two Germans, Matthias Schleiden, a botanist, and Theodor Schwann, a zoologist, independently advanced the radical "cell theory." Although they did not originate the idea (the French biologist René J. H. Dutrochet and the German naturalist and philosopher Lorenz Oken first asserted that cells were the fundamental structural units of life), Schleiden and Schwann elevated it to one of the most important theories in science. The fundamental units of life, they said, were not fibers and vessels or other gross anatomical structures. They were microscopic cells, each of which was a living entity of its own. The term "cell" had arisen two centuries earlier when the English scientist Robert Hooke looked through an early microscope at a thin slice of cork and saw rectangular chambers that reminded him of monks' cells in a monastery. He called the microscopic voids "cells" (Figure 1.1). Hooke's cells, it later turned out, were the empty spaces inside the dead and dried cell walls of a cork tree. Schleiden and Schwann revived the term for microscopic chambers of living matter, applying it not to Hooke's empty spaces but to the packages of living stuff within the spaces.

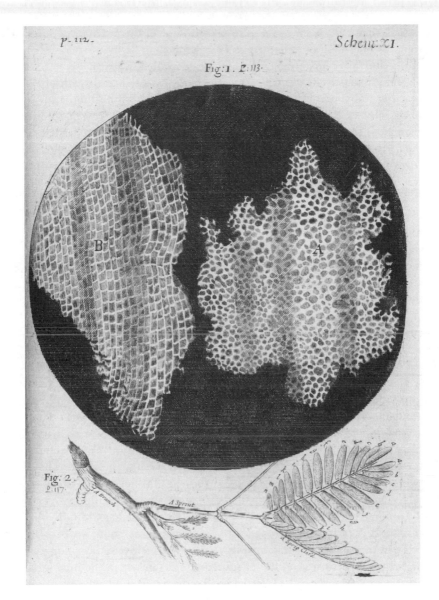

**Figure 1.1** When the seventeenth-century English scientist Robert Hooke peered through an early microscope at a slice of cork, he saw vacant chambers that reminded him of the cubicles inhabited by monks, which were called cells. So he gave that name to the spaces that a later generation of researchers would come to regard as anything but empty. (From Robert Hooke, *Micrographia* [London, 1667]. Photo courtesy of the Chapin Library of Rare Books, Williams College)

Schleiden even went so far as to perceive one of the fascinating philosophical questions of cell theory: If a cell is a unit of life, then is a multicellular organism really a colony of one-celled individuals?

Some plants and animals, after all, live part of their lives as free-roaming individual cells and only later congregate by the thousands to form colonies that look like multicelled organisms. One well-studied example is the cellular slime mold called *Dictyostelium,* which has one of the more remarkable life cycles in nature. Individual cells of *Dictyostelium* roam the forest floor as free-living amoebas, but if food supplies turn scarce, they congregate in huge masses and assemble themselves by the thousands into a much larger multicelled organism that looks and acts like a slug. The slug creeps about for a while, then settles down and metamorphoses into something that looks more like a fungus. The blob of cells sprouts a thin stalk that projects upward perhaps a quarter of an inch. Then other cells of the slug climb up the stalk and organize themselves into a ball balanced on top. The outer cells harden into a shell; the inner cells shrivel into dustlike spores. Eventually the shell cracks open and the spores are scattered to the wind. Spores that land in wet places change into amoebas and slither off to begin the cycle anew.

Another challenge to Schleiden's definition of an organism is the living sea sponge. Push one through a fine-mesh sieve and its cells will separate from one another, turning clear aquarium water into a thick, cloudy liquid, like pea soup. Wait a few hours, however, and the cells will gradually find one another, stick together, and reassemble themselves into a whole sponge. Although sponges come in many species, each with a distinctive appearance, the individual sponge cells will invariably rebuild the correct architecture for its species. In fact, the disaggregated cells of two different sponge species can be mixed, and the cells will sort themselves out and reassemble only with their own kind, re-creating sponges of the original two species. In other words, contained within each individual cell is a part of the "knowledge" of how its species is built.

Schwann compared an animal made of cells to a hive of bees, suggesting that each hive was a kind of superorganism in which the individual cells, or individual bees, had a separate identity and could have an autonomous existence. In the early days of cell theory, incidentally, philosophical speculation about the possible colonial nature of the human body quickly found its way into debates in political philosophy about the role of the individual in relation to society as a whole.

One error in Schwann's thinking was his idea that cells arose anew out of some generalized body fluid. In the 1850s a major advance in cell theory laid this idea to rest. Rudolf Virchow, a German pathologist and a strong advocate of the cell theory, suggested instead that all cells are formed by the division of preexisting cells. It was a crucial modification of cell theory, for it opened the door to an understanding of how organisms develop, growing from one cell to many. Virchow also advanced the concept of the cellular basis of disease, the view that all disease arises because of afflictions within or among cells. Virchow was an early proponent of the idea that life is an essentially mechanical process—that it can be explained entirely by the workings of the laws of physics and chemistry and without any need of the vitalists' supernatural forces.

In their battle with the vitalists, the mechanists did not mind stating their views in provocative terms. "A man is what he eats," was one slogan, noting that every atom of the human body has been extracted from food. (To be totally accurate, the mechanists should have included water and air as sources of atoms in the human body.) Not even the stuff of the mind, the mechanists argued, had a mystical source. "The brain," some liked to say, "secretes thought as a kidney secretes urine."

Over the next few decades Virchow's ideas encouraged more detailed study of cell division. Improved microscopes and better methods of staining cells to make their normally colorless insides visible led to a slowly improving knowledge of both the internal architecture of the cell and the events of cell division—the process that gave rise to all existing cells. Still, much would remain unknown into the late twentieth century. By 1882 Walther Flemming, yet another German, produced the first detailed description of cell division, including the central phenomenon of mitosis, the creation of two identical sets of genetic material (that is, two sets of chromosomes), each an exact copy of the one set in the parent cell.

By 1900 cell theory had led to a coherent view of how a complex organism arises. It was clear that the organism begins as a single cell, formed at conception by union of the father's sperm and the mother's egg, or ovum. It was known that sperm and egg each carried a set of hereditary factors, or genes, from one parent. And it was understood that the combined genetic endowments were duplicated during each round of cell division and one complete set of genes passed on to each new cell.

In 1912 the mechanist movement received a major boost from a German-born biologist named Jacques Loeb, who had immigrated to the United States and began working summers at the Marine Biological Laboratory in the Cape Cod fishing village of Woods Hole, Massachusetts. In 1912 Loeb published a landmark book entitled *The Mechanistic Conception of Life.* In it he described his own experiments on sea urchin eggs. Loeb had found a way to remove the eggs from a female urchin and make them start their embryonic development, just as if they had been fertilized by sperm—but without sperm. Loeb found that a simple dose of certain lifeless chemicals would launch one of the most dramatic phenomena in all of biology, the development of an organism from a single cell. Loeb offered his research findings as confirmation of the mechanistic view.

Nowadays the procedure is quite routine and is performed even in high school biology classes to demonstrate the early stages of embryonic development. But in Loeb's day word of his experiments captivated the public. Newspaper headlines, reflecting the naiveté of early science writing, virtually claimed that Loeb could create life in a test tube. Some compared it to reproducing "virgin birth" in the laboratory, starting living cells on a course that had long been regarded as miraculous. Loeb's experiments, in a seaside laboratory and on marine organisms, became so mixed up in the popular mind with the allegedly mysterious powers of the sea that unmarried women were advised not to bathe in the ocean. Childless couples, on the other hand, rushed hopefully to beach resorts. Loeb even received letters from desperate couples asking him to give them a child.

As the inheritor of a German tradition in which great thinkers in science felt free to pontificate on all manner of social and political issues, Loeb wrote and spoke widely. Like Virchow he espoused the society of cells as a model of social cooperation. Loeb hobnobbed with the literary figures of his day—Thorstein Veblen, H. L. Mencken, even the young Gertrude Stein, who studied marine biology at the Woods Hole lab for a summer. Sinclair Lewis made Loeb the model for Max Gottlieb, the wise scientific mentor of Martin Arrowsmith, the idealistic hero of his famous book *Arrowsmith*.

Loeb's thinking was also influenced by the rediscovery in the early twentieth century of Gregor Mendel's lost writings on the breeding of garden peas in his monastery garden. In the mid-nineteenth century, Mendel had discovered powerful evidence that hereditary traits are passed from one generation to the next in discrete units, parcels of heredity that today we call genes. But the monk's publications were little noticed nor long remembered. They were found again, however, in the opening years of the twentieth century and seized upon by the mechanists as evidence that molecules—which are, of course, discrete units—govern heredity. In 1911 a prescient Jacques Loeb grasped the significance of Mendel's findings and knew from more recent experiments that at every cell division the chromosomes were duplicated and parceled out in equal numbers to the daughter cells. The biochemist's chief job, Loeb said, was to find "the chemical substances in the chromosomes which are responsible for the hereditary transmission of a quality." Loeb was calling for the discovery of the structure of DNA and the genetic code.

It took longer than Loeb probably anticipated, but in the next half-century biochemists would do exactly that. By applying a purely reductionist approach to the study of life, they would vindicate the mechanist view spectacularly with the discovery of the double helix of DNA and the cracking of the genetic code. While the biochemists were pursuing a largely molecular approach to life, a parallel branch of research, cell biology, was studying the forms and functions of whole cells. Eventually, in the 1970s and 1980s, the molecular approach and the cellular approach would merge into a relatively unified, and utterly mechanist, approach to the study of life itself.

But Loeb went even further. He was confident that the mechanics of life would prove simple enough that it should be possible to create life in the laboratory. "We must either succeed in producing living matter artificially, or we must find the reasons why this is impossible," Loeb wrote. Today, as we shall see, many of the phenomena of life—many of the events within cells that, together, make for life—can be made to happen spontaneously under artificial conditions in the test tube. Whether that would have satisfied Loeb is not clear. Most biologists today think it will be a very long time before anything like an artificial cell could be synthesized.

Loeb even saw ethics as the product of the mechanical processes of life:

If our existence is based on the play of blind forces and only a matter of chance; if we ourselves are only chemical mechanisms, how can there be an ethics for us? The answer is that our instincts are the root of our ethics and that the instincts are just as hereditary as is the form of our body. We eat, drink and reproduce not because mankind has reached an agreement that this is desirable, but because, machine-like, we are compelled to do so. The mother loves and cares for her children not because metaphysics had the idea that this was desirable, but because the instinct of taking care of the young is inherited. We struggle for justice and truth since we are instinctively compelled to see our fellow beings happy.

The adult human body, according to current estimates, is made up of sixty trillion cells—eleven thousand times as many units of life as there are human beings on Earth. The body also contains various nonliving materials, such as hair, fingernails, and the hard part of bone and tooth, all produced by cells. The most visible part of the body, the skin, is made up of the thickly matted, fibrous "skeletons" of dead skin cells. Even the fat of "middle-aged spread" resides within special fat-storage cells that can swell to huge proportions.

Human cells vary greatly in size, from the tiny red blood cell at 1/25,000th of an inch (0.00004 inch) across (Figure 1.2) to a typical diameter ten times larger of 1/2,500th of an inch (0.0004 inch) for a kidney cell or a liver cell to the gigantic muscle cells that can be thin filaments a few inches long. The record holders are the nerve cells that begin at the base of the spine and run all the way to the tip of the big toe—a distance of several feet.

In the human body there are about 200 different kinds of cells with different shapes and jobs, but all are similar in basic structure and internal workings. And, although each cell is only a minor constituent of a multicelled organism, much of modern medical progress can be traced to one startling fact: Many kinds of cells can be removed from the body and

**Figure 1.2**  Doughnut-shaped red blood cells on the point of an ordinary pin. Small as human cells are, these erythrocytes (as they are also called) are about one-tenth the diameter of a more typical-size cell. Red blood cells measure about 1/25,000 inch across. Their function is to take up oxygen from the lungs and carry it to cells throughout the body. (American Society for Cell Biology)

*in a dish*

allowed to live as independent organisms. This is the fact that makes possible the cell cultures at ATCC and in scores of other research laboratories around the world. The discovery of the potential independence of our cells confirms an early speculation of Schleiden's that cells lead what he called "double lives"—their own and that of the organism of which they are a part.

*cultured cells*

Snip a tiny chunk of tissue from almost any part of your body—a piece of skin, for example, that would fit inside this o. Treat it briefly with protein-digesting enzymes that break the proteins holding the cells together. Drop the disaggregated cells (there would be thousands in a piece this size) into a teaspoonful of nutrient fluid. The cells that once were loyal, hard-working members of a vast organization of trillions of cells will revert to a way of life much like that of their evolutionary ancestors, the one-celled protozoans. Human cells retain a primordial capacity for independent life, crawling about their culture dishes like amoebas on a pond bottom, feeding on nutrients in the water, reproducing by cell division. Many cells even take on an amoeboid form, becoming sprawling blobs that constantly change shape as they undulate across the bottom of their dishes (Figure 1.3).

Even within the human body there are cells that live rather like protozoans. Certain white blood cells, for example, are part of the immune system and flow as free individual cells with the blood. But they can detect bacterial infections and exit the circulatory system to enter affected tissues, where they crawl around to catch and eat the invaders. Like amoebas, they simply engulf the bacteria and digest them.

One specialized member of the immune system is called the "natural killer cell." Millions of them roam the body, searching for other cells that have turned cancerous. Once the killer cell finds its prey, it presses close and exudes a substance that kills the cell. This newly emerging understanding of natural killer cells, incidentally, is leading to new ideas on the prevention, cause, and cure of cancer. Some researchers suspect cells frequently turn cancerous but are usually killed before they can proliferate into tumors. One reason tumors arise, then, may be defective killer cells, so some researchers are looking for ways to cure cancer by boosting the number of killer cells in the body.

Still another kind of amoebalike cell in the body inhabits bone. When bones are growing, or healing after a fracture, these cells crawl through the hollow spaces something like a snail that leaves a slime trail. The bone

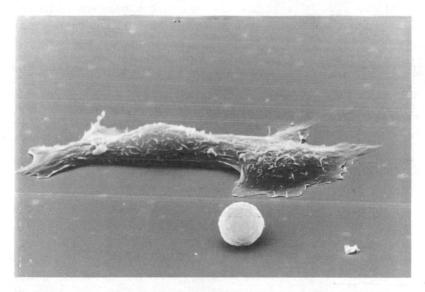

**Figure 1.3** Crawling across the bottom of a culture dish, a disembodied human prostate cell behaves much like its evolutionary predecessors, the one-celled protozoans. The ball is another cell that has rounded up in preparation for cell division. After splitting in two, each daughter cell will spread again.

*motility*

cell's slime, however, is a substance that hardens into the mineral part of bone, gradually building the bone's thickness. Other motile bone cells do the opposite, taking up previously deposited bone by dissolving it in their path. Like sculptors, the two kinds of cells work in concert, removing bone here and adding bone there, to remodel the tiny bones of a baby into the big ones of an adult. When bones break, these cells receive a special signal that spurs them to knit the fragments together. Even if the bone fragments grow together in a crooked form, the bone sculptors will continue reshaping the affected area until it becomes normal again.

Skin cells also become motile to heal wounds. Imagine a cut finger or a scraped knee. New skin doesn't grow only at the edges of the wound, repairing it progressively from the edges toward the middle. Special proto-skin cells crawl from the bottom layer of the epidermis and onto the exposed wound surface, where they multiply rapidly and distribute themselves in a thin layer over the entire damaged area. Once the cells have formed a complete layer, they begin dividing with a new result. The daughter cells change form and rebuild the skin's normally multilayered structure.

Surely the most spectacularly motile human cells are sperm. Like many of their primordial ancestors in pond water, they swim by lashing their taillike flagella.

Because individual cells are virtual organisms in their own right, some biologists maintain the view of Schleiden and Schwann and think of each multicelled organism, including each human being, as a community of organisms, a huge colony of extraordinarily selfless citizens, each forsaking independent existence for the good of the colony. The kidney cell patiently resides deep within its own province, performing its specialized job of removing waste materials from the blood. The skin cell clings tightly to its neighbors to protect the body against infection and desiccation, dutifully staying put—unless, of course, an injury requires it to mobilize its healing potentials. The human body is a republic of cells, a society of discrete living beings who have, for the good of the society as a whole, sacrificed their individual freedoms.

*Cell death*

So profound is this sacrifice that cells have gone so far as to sign their own death warrants and hand them over to their neighbor cells. Then, should a cell's continued existence become a threat to its society, the neighbors may execute the warrant and leave the cell no choice but to commit suicide. Upon command of its neighbors, the hapless cell initiates a genetic program that carries out its suicide in ritual precision. Although evidence of this extraordinary form of natural cell death has been known for decades, only since the early 1990s have cell biologists accumulated evidence that this "programmed cell death" is an integral part of multicellular life.

This phenomenon, sometimes called *apoptosis* (from the Greek words that refer to leaves falling off a tree in autumn), is in fact the way "natural killer cells" destroy cells that have turned cancerous. It also plays a key role in embryonic development, performing the cellular equivalent of removing the scaffolding after construction is complete. For example, during the fifth week of human embryonic life, the hands are flat paddles with no distinct fingers. Then, during the sixth week, four waves of programmed cell death roll across the paddle, removing the cells between the fingers. Even in adult life, the death program continues as millions of cells kill themselves every minute in the human body. We'll return to this extraordinary process in detail in later chapters.

Of course, cells have also gained something in the cooperative bargain. By banding together they create an environment that is far more stable and nurturing for each individual than the outside world can provide. By collaborating in various specialized jobs, the cells that make up a human body create huge systems that maintain ideal temperature, protect against drying out, and provide ample supplies of oxygen and nutrients.

For all the evident differences among specialized cell types, one of the most profound insights to emerge from modern cell biology is that all cells, even those from species as different as a yeast and a human, carry out the same, fundamental housekeeping chores of life in exactly the same way. A human brain cell, for example, is not more complicated than a one-celled creature inhabiting pond scum. Indeed, all the fundamental processes needed to sustain life are identical from the amoeba cell to the human cell, just as the fundamental processes of a sedan and a station wagon are identical. The differences are in the added special functions and shapes.

Though animal cells were confined to communal existence billions of years ago when multicelled animals evolved from one-celled ancestors, liberation did not come until 1885 when Wilhelm Roux, still another of the Germans who dominated early basic biology, discovered that a patch of cells removed from a whole organism (from a chicken embryo in this case) could live for several days in a dish of saline solution. Further success in culturing individual cells came in 1907 when the American biologist Ross G. Harrison of Yale University found a way to maintain frog nerve cells in a dish. He found that isolated nerve cells in his dish even sprouted thin filaments that reached out and made contacts with other nerve cells, as if they were trying to form a primitive nervous system.

Further progress in cell culture methods came slowly, as researchers sought mainly for a better understanding of the environmental conditions and nutrients that cells need to live and multiply. It was not until the early 1950s that cell culture became fairly routine. It was also at that time that the cancerous cells of Henrietta Lacks were removed and established in culture as HeLa cells.

The main thing cultured cells need is to be bathed in a fluid that contains some food—mainly sugar, mineral salts, some vitamins, and some amino acids, which are the building blocks of proteins. The fluid should both supply oxygen and take up the waste product carbon dioxide. When kept warm in such a fluid, the tiny blobs of life thrive. They take in the nutrients, reprocess them to grow and to carry out their specialized functions, and then excrete the waste products. To avoid programmed cell death, cultured cells also must be given a supply of "growth factors" produced by other cells.

As the cells creep about in their plastic dishes, they thrust out a wide, ruffled edge that seems to search the space ahead for cues as to which way to go. Parts of the leading edge stick down to the surface and the cell's body appears to drag its hind parts along. When two cells bump, they cringe, shrink back and quietly slither off in new directions.

Periodically the cells cease their travels, let go of all attachments to the surface, pull themselves into a round ball, and undergo perhaps the most dramatic of life's many astonishing phenomena—cell division. The cell breaks down some of its old internal structures, recycling the components to assemble new ones. The cell's genetic endowment, made of the chemical DNA, is copied into a duplicate set of genes. Then the old cell pinches in half to become two young cells. Through this routine but utterly astonishing process of mitosis, many cells are reincarnated without ever dying.

After several generations of doubling in laboratory containers, the cells become too crowded, and their human keepers must come to their aid. Some of the cells are removed to found new colonies in new dishes of nutritive fluid. The amount removed is critical because different types of cells have different requirements. Some are content as loners, and only a few are needed to start a new population in a new bottle. Others, as one biologist put it, "need to have their friends around. They die if they get too lonely."

Many cells living in culture remember their old lives as parts of tissues in larger organisms (Figure 1.4). Skin cells, for example, will multiply until they form a sheet covering the bottom of their container, just as they used to make skin in their native habitat. Under certain conditions, the layer of skin cells will even develop into proper, multilayered skin—a human tissue assembling itself right in a dish. Today, researchers are learning to control this process in the hope of creating skin farms in the laboratory as sources of human skin that could be grafted onto burn patients who have lost some of their own. Also in the culture dish, cells called fibroblasts will keep on secreting collagen, the tough fibrous protein they once made to give strength to skin and other tissues. Cells of the female breast will manufacture and secrete milk protein. Muscle cells will sometimes weld themselves into large fibers that spontaneously begin twitching in the dish. When cells of the heart muscle do this, they begin beating rhythmically. Even disembodied nerve cells, as Harrison discovered in 1907, will sprout long, thin filaments of living tissue that reach across the surface of the dish, as if seeking another cell to deliver an electrochemical message. And, as recently as 1990, researchers discovered that human brain cells can be induced to multiply in a dish. The cells link up to form synapses, the connections that relay signals from one nerve cell to another. The little knots of brain cells behave as if they were assembling themselves into a primitive brain.

For decades, the only way to maintain human cell cultures was to keep them in incubators heated and thermostatically controlled to maintain a temperature of 98.6 degrees Fahrenheit—human body temperature. Some cell cultures removed from a fully developed human or other animal would proliferate for a number of rounds of cell division, and then mysteriously, the cells would die. Somehow they lost the ability to keep dividing. Cells taken from cancers, on the other hand, proved apparently immortal and would keep on living and dividing indefinitely. The very property that made them deadly to their former hosts now made them invaluable to cell biologists.

Then in 1949 Alan Parkes, a British biologist, developed an entirely new way to maintain cell lines—freezing them. Researchers had been attempting this for decades—especially as a way of storing the semen of prize bulls so that it would be available for artificial insemination after the animal died. Many experiments simply killed the cells. Some found that a few cells revived on thawing but that they were sickly. Finally Parkes found that if he added some antifreeze to the nutrient bath and then cooled the culture very slowly, the cells would enter limbo and could revive on warming. For this antifreeze, Parkes found success with glycerol, a clear, syrupy liquid better known to many as glycerin, which is also used in automobile antifreeze. Nowadays a better substitute is dimethyl sulfoxide, or DMSO (a clear liquid by-product of wood pulp manufacture that has many uses from paint remover to an anti-inflammatory drug).

Either way, however, the antifreeze is toxic to cells, so once they have been treated with it, they cannot be allowed to continue in the active state of life for too long. A batch of cells is placed in a vial, or ampoule, along with a watery solution that is 95 percent nutrient medium and 5 percent antifreeze. At ATCC the ampoule is put in a special freezer that cools it from room temperature to 112 degrees below zero over an hour. Faster freezing will kill cells. In many labs the vials are simply placed inside a block of thick styrofoam insulation and put directly in a freezer that

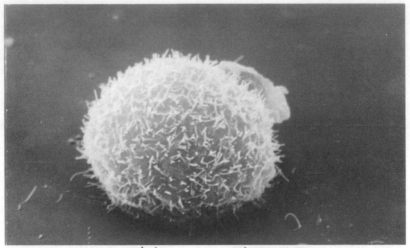

Figure 1.4a    *prostate – ready to divide*

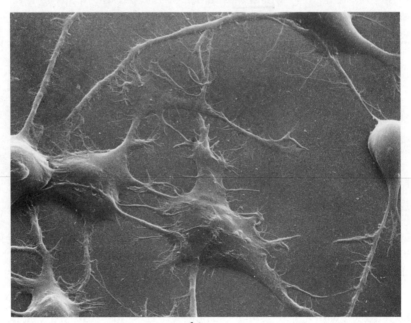

Figure 1.4b    *Nerve cells*

**Figure 1.4**  Cells take many forms in culture: (a) the splendid isolation of a rounded-up human prostate cell about to divide, its surface covered with hairlike projections; (b) nerve cells reaching out to communicate with neighbors, forming a rudimentary nervous system; (c) epithelial cells that have spread and joined at the margins to form a skinlike sheet; (d) a colony of prostate cells beginning to become crowded. (Images b, c, and d courtesy of Guenter Albrecht-Buehler)

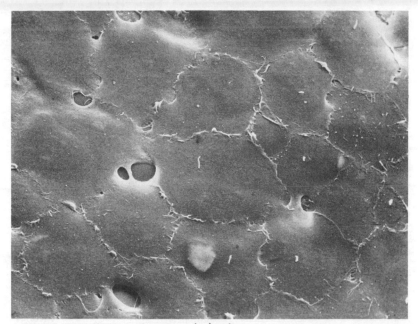

Figure 1.4c    *epithelial*

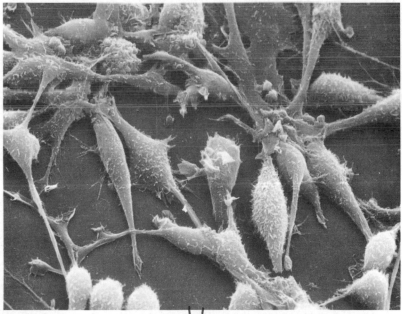

Figure 1.4d    *prostate*

maintains this temperature. The insulation slows the rate of heat loss to just the right speed.

What happens is that the water outside the cells freezes before the water inside the cells. This is because all the molecules normally present inside the cell lower the freezing point. The antifreeze keeps the water crystals outside the cell from becoming large enough to pierce the cell membranes. As the outside water slowly crystallizes, however, the liquid water inside the cell is drawn out, diffusing through pores in the cell's membrane and joining the ice forming outside the cell. The cell shrinks, literally deflated by the loss of water. If conditions are right, virtually all of the water will have left the cell before the inside becomes cold enough that it would freeze water into crystals big enough to rupture internal membranes.

As the temperature in the vial falls, all the metabolic processes of each cell—the hundreds of simultaneous but different biochemical reactions that, in the aggregate, amount to life—become slower and slower. When the temperature gets cold enough, they stop altogether. Like a microscopic bear going into hibernation (except that bears keep metabolizing at a slow rate), the cells gradually relinquish the usual textbook criteria for being alive. All that remains of these life-forms is the form itself—the architectural structures that, when warm, are unquestionably alive.

Are frozen cells dead? They do not eat or move. They do not even carry out the slightest form of metabolism. By the usual definitions of life, a frozen cell is not living. And yet, when the cells are warmed, they resume all the activities of life. When that happens, is life suddenly created anew? Does life arise spontaneously from nonliving matter? Is it resurrection?

Many biologists answer that this is just a semantic trap, and that frozen cells should be considered neither alive nor dead but in a state of suspended animation. The key point is that what remains in the frozen cell is the integrity of its structures, from the unbroken cell membrane to intact forms of all internal architecture and molecules.

So long as an organism maintains its structural integrity, one can argue that it is alive. In this view, life can persist without anything happening, much as a car is still a motor vehicle when it is not running. The crucial factor in both cases is whether the component structures are present and linked in the right ways so that they could interact if given the right conditions.

The convenience of cold storage was immediately evident to officials at the American Type Culture Collection (ATCC), which had been started in 1925 to supply scientists from its reference collections of all forms of life, including bacteria, fungi, protozoans, and plants. It's cell bank was put on ice. Or, more properly, it was put on liquid nitrogen, the clear, colorless, odorless liquid (made in factories by chilling nitrogen gas until it condenses) that always stays a frosty 321 degrees below zero. As long as liquid nitrogen is kept well insulated, it stays cold with no need for motors or compressors.

It is liquid nitrogen that chills ATCC's stainless steel tanks and keeps their scientifically precious contents well within that mysterious state of biological limbo. Each tank is about halfway full of liquid nitrogen. Inside the tank some ampoules, clipped into metal holders, hang from a rack submerged in liquid nitrogen. Some hang from a rack in the nitrogen vapor just above, not as cold, but at 211 degrees below zero, still quite frozen.

"Say a researcher at a laboratory somewhere is working on a disease that involves a certain kind of cell in the body," Hay explains as he stands in the large first-floor room that ATCC researchers call the Tank Farm. More than forty stainless steel freezer tanks crowd the room, thirty-two of them devoted to cell culture and others to microbes and protozoans. If ATCC has what the researcher wants, he or she places an order. Unless the appropriate cells have already been thawed and are growing in special cell culture bottles in an incubator upstairs, a technician will have to revive the frozen cells waiting in an ampoule.

From the inert microscopic forms, unmoved and unmoving for perhaps many years, will emerge the long-suppressed pulsations of life. This simple phenomenon, this "resurrection," repeated in laboratories around the world hundreds of times a day, poses a profound challenge to the ancient view of life as a mystical process. On the other hand, "resurrection" is perfectly consistent with the view of modern cell and molecular biology that life is a mechanical process that is, in principle, no harder to understand than an automobile or a computer.

In place of the animating "vital force" of 150 years ago, modern biology confirms the view that all the phenomena that together constitute life can be understood in the terms of chemistry and physics. The frozen cells possess the right chemicals in the right combinations and in the right structural arrangements to live. The cold temperatures simply deny the cells the thermal energy needed for the chemical reactions to proceed. Like a car battery in a Minnesota January, the chemistry simply won't go when it's too cold. But add a little energy in the form of heat, and the chemistry happens.

Science is still a long way from being able to explain every detail of how life works, but in recent years basic biology laboratories have been founts of astonishing new knowledge. The field is racing many times faster than it did a generation ago when James Watson and Francis Crick figured out the double helix structure of DNA. Almost every day, some new detail of life's machinery is revealed.

Cell and molecular biologists can now explain many of life's most intimate workings as the results of purely nonliving events—interactions among atoms and molecules no more mysterious, though often far more complex and wondrous, than the crystallization of water molecules into snowflakes. As beautiful and as mathematically precise and as predictable as snowflakes, their formation is obviously no miracle. Crystallization of water is understood down to the last atom and can be produced in the laboratory at will or, obviously, in the kitchen refrigerator. It happens because certain molecules have the innate property of organizing themselves into predictable arrangements entirely without guidance from an outside intelligence. Many atoms or molecules will, when put into the right environment, link together in regular patterns to form crystals. Some are cubic crystals, some are hexagonal. Others take other forms. The key point is that no outside force need guide the events. The guidance is implicit in the molecules themselves. Certain molecules assemble themselves into specific, predictable structures because their shapes, their own physics and chemistry, will allow them to assemble only in those patterns.

This concept of self-assembly is fundamental in today's cell and molecular biology. Under the right conditions, atoms or small molecules automatically link—like water molecules assembling themselves into snowflakes—to form larger structures of absolutely predictable shape. Life's chemistry generally involves very large, highly complex molecules such as the proteins, but these too assemble themselves into still larger, predictable structures. Incidentally, although many people think of protein as something good to eat, cell and molecular biologists think of proteins as many different kinds of molecules, some of which act as little machines that carry out specific actions in the cell or as units of structure, such as bricks or girders, that can combine to make the architecture of the cell. A typical cell may contain several thousand different types of protein molecules.

The discovery of self-assembly explains many events in cells that once seemed utterly mysterious. It is akin to learning that the steel skeleton of a building will assemble spontaneously once a load of girders is dumped at a construction site. The genes cause a load of protein molecules to be synthesized in the cell, but the proteins, needing no further control, spontaneously assemble themselves into larger structures. The plan of the final result is implicit in the structure of the components (Figure 1.5).

Laboratory experiments have shown that self-assembly is even more amazing. Take batches of several different kinds of subunits and mix

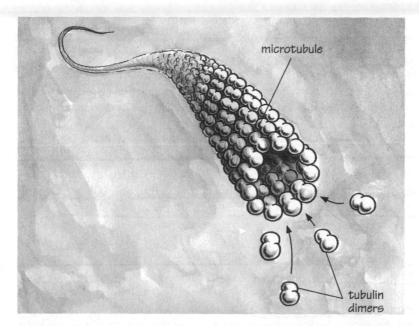

Figure 1.5a

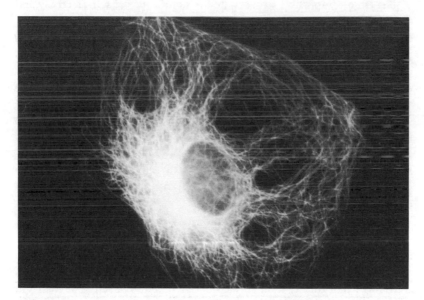

Figure 1.5b

**Figure 1.5**   (a) Like magic bricks that assemble themselves into a wall, molecules of tubulin (a protein that all cells make) will spontaneously link, if conditions are right, to form long, thin tubes, called microtubules. These filamentous structures function as tracks over which motor molecules transport containerized cargo throughout cells. (b) The microtubule network in one cell.

them together in what would seem to be a hopeless stew of hundreds of different substances. The molecules sort themselves out flawlessly and assemble perfectly. It is as if the load of girders were mixed up with a load of bricks and pipes and boards and plaster and nails and tiles and still the girders sorted themselves out and assembled spontaneously. And, at the same time, so did the bricks and boards and all the other components of the building.

Similar phenomena, as Jacques Loeb prophesied at the beginning of the twentieth century, lie at the heart of all life. Molecules of a given shape and composition possess the power not only to link themselves into larger structures but to act on entirely different molecules, causing them to break apart in specific ways or to combine with still other molecules in predictable ways. The molecules that act upon other molecules, for example, are intracellular mechanics called enzymes. As catalysts, enzymes make things happen chemically but remain separate from the result, able to work their effects again. An average cell contains many hundreds of different kinds of enzymes, each built to perform a specific job. Like their cousins, the structural proteins, enzyme proteins do their work automatically because of their shapes and the inherent chemical proclivities of their components.

It is almost impossible to overstate the significance of self-assembly and the inherent powers of enzymes, for these processes have profound philosophical implications. As the French molecular biologist Jacques Monod put it in his book *Chance and Necessity,* ". . . an internal, autonomous determinism guarantees the formation of the extremely complex structures of living beings." Although Monod published in 1970 at what now seems to be the Paleolithic era of molecular biology and few examples of self-assembly were available, he has turned out to be entirely right. So too have Loeb's very similar pronouncements more than half a century earlier. Monod died in 1976, but molecular and cell biologists continue to cite his book as an outstanding exposition of a philosophy that accepts a mechanistic view of life and celebrates its resultant glory—the entire living world of Earth.

Monod also saw that this view could resolve an old debate between two schools of thought about the forces that molded living organisms, including human beings. One group was the preformationists, adherents of an ancient doctrine, who claimed that the anatomical details of every person were formed at the creation and simply bloomed, in a sense, at the right time. The oldest version of preformationism even held that every sperm carried a miniature human being, called a homunculus, in whose sperm there were even tinier humans and so on. The rivals were the epigeneticists, disciples of a newer school that accepted the role of genes (their existence had become evident, though not understood, early in the twentieth century). The epigeneticists insisted that genes dictated only the structure of certain molecules. They argued that molecules needed some additional guidance—which they called epigenetic and which many regarded as supernatural—before the molecules could be assembled into larger structures such as cells and whole organisms.

The discovery of the phenomenon of self-assembly, Monod wrote, should end the debate by showing that both are, in a sense, right but that neither need invoke phenomena beyond those natural processes that can be understood by science. "No preformed and complete structure pre-existed anywhere; but the architectural plan for it was present in its very constituents. It can therefore come into being spontaneously and autonomously, without outside help and without the injection of additional information. The necessary information was present, but unexpressed, in the constituents. The epigenetic building of a structure is not a creation; it is a revelation."

Monod's book was deeply disturbing to many because it asserted that no event in the life of a cell or, indeed, in the life of a whole human body, was the result of any supernatural guiding hand. Today, although everything that happens in the living cell still cannot be explained on the basis of known chemical reactions, the progress has been so spectacular that it would be hard to find a cell biologist today who thinks that the unexplained processes will not someday be fully understood.

"The secret of life is not a secret anymore," says Duke University cell biologist Harold Erickson. "We've known for twenty or thirty years now that life is not more mysterious than the chemical reactions on which it is based. There's an incredibly complex set of chemical reactions, but they're all logical and understandable. We don't yet understand them all but we do understand a lot of them and it's not hard to see that eventually we should know them all."

Tom Pollard, a cell biologist who heads the famed Salk Institute in La Jolla, California, and a former president of the American Society for Cell Biology, questions whether nonscientists are prepared to accept the idea that life is just so much chemistry and physics: "What molecular biologists have believed for two generations is now generally regarded as proved beyond any doubt. Life is entirely the result of physics and chemistry inside cells and among cells. I wonder whether the general public is prepared to sign on?"

The old, mystical view of life denied any opportunity for expression of the most wonderful aspect of human life, the intellect. But apply the intellect to the rational contemplation of modern cell and molecular biology and what emerges is the awareness, both chilling and inspiring, that the human body is a consummately wondrous assemblage of cells that are each machines. So are all forms of life. Although some religious people complain that this view of life removes any role for a supernatural creator, it need not do so. Just as most of the world's established religions accept evolution as the process by which their deity created the world, they also may come to see the new cellular and molecular biology as revealing the mechanisms by which that creation made life possible.

## CHAPTER 2

Every biologist has seen it and marveled. Some watch it for hours, spell-bound. Take an individual living cell—a human skin cell, for example—and put it under a powerful microscope. What comes into view is a far cry from the vague "protoplasm," the so-called living jelly, "the stuff of life" that was taught in high school biology courses a generation ago.

"There's a whole world buzzing in there," says Robert D. Goldman, a cell biologist at Northwestern University's medical school in downtown Chicago. "You can see all kinds of different things. Some of them just sit there and kind of jiggle but lots of them are traveling, going somewhere."

Goldman, who gets visibly excited when he talks about cells and microscopes, spies on the inner workings of individual cells both by peering directly into a microscope and also by using a special video camera that is sensitive to differences in light levels the eye cannot see. The camera looks into the eyepiece and relays what it sees through a battery of electronic devices that pick out the subtle contrasts and emphasize them. It displays the enhanced image on a video screen. Unlike an electron microscope, which requires cells to be killed first, the video-enhanced light microscope lets cells live and carry out their internal processes.

The image is nothing like the usual cell diagram that shows a few lonely objects drifting in an otherwise empty intracellular sea. Real cells are jammed full of objects. Their insides are more crowded with components than the inside of a computer or what's under the hood of a new car. There are thousands of specialized structures, each with a job to do. And most of the structures are in motion. If the human body is a collection of many internal organs, each with its own functions and behaviors, the cell is a collection of many little organs, each also with its own form and function. In cells the little organs are called organelles.

On the video screen thousands of tiny shapes, each a hollow vessel of chemicals, jostle about. Many glide in straight or curved lines—some smoothly, some in fits and starts. Small spheres jiggle in place or lurch ahead. Dark, sausage-shaped objects loom into sight, turn a corner on some unseen road, and slither back down out of the microscope's plane of focus. Wormlike structures undulate across the field of view (Figure 2.1).

Off to one side of the cell sits the most obvious organelle of all, the gigantic brooding dome of the nucleus, repository of the genetic blueprints that govern the cell's forms and functions. Still invisible, even with video-enhanced microscopy, is the ceaseless movement of molecules of all shapes and sizes entering and exiting the nucleus through special portals. Among those emerging from the nucleus are long threads of molecules that, like messengers from the king, carry instructions issued by the chromosomes inside. Within minutes the coded commands will be carried out by one or another of the hundreds of organelles in the cell.

Life at its most fundamental is dynamic. Even among sedentary organisms such as a coral or a tree, motion abounds within their cells. Inside each of the estimated 60 trillion cells that make a human body—from the strongly joined cells that are skin to the obscure working cells deep within

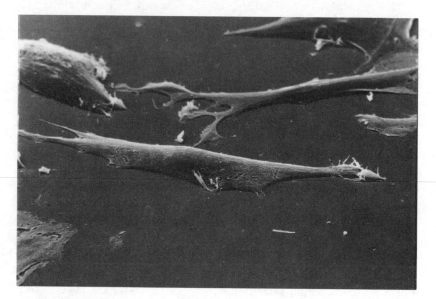

**Figure 2.1a**

**Figure 2.1**   Three ways to look at cells. (a) Only surface contours show in this image produced by a scanning electron microscope. (b) With a light microscope, internal structures are evident, the largest being the big, oval shape that is the nucleus. Video images are like this. (c) The most detailed view emerges from the transmission electron microscope. In this highly magnified image, only one edge of the nucleus appears (lower right), the nuclear membrane angling upward from the lower left. Also visible are several organelles that will be examined in later chapters: Golgi, the endoplasmic reticulum, ribosomes and lysosomes.

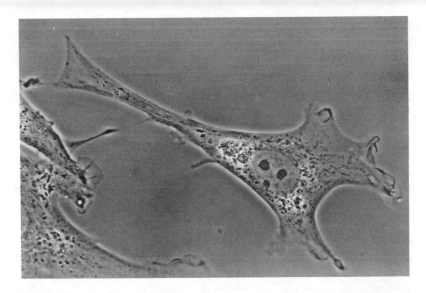

Figure 2.1b

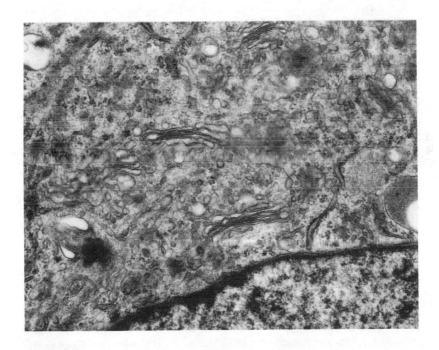

Figure 2.1c

the liver to the freely motile cells that swim the bloodstream—life is an unresting machination of movements. Indeed, autonomous motion has classically been one of the most fundamental attributes of life. The ability of a hunk of matter to pick itself up and travel in a purposeful way struck ancient philosophers as of profound significance—so profound that they could not attribute it to mortal flesh. The ancients could not see within cells, but if they had, they would have found even more frenzied examples of motion in the ceaseless concert of thousands of simultaneous chemical and physical reactions that go on.

To anyone watching this magnified spectacle, it can seem as chaotic as a beehive. Yet, just as a beehive can be understood through a knowledge of the separate roles of its queen, workers, and drones, so the phenomenon of life begins to make sense through an analysis of the organelles within a cell.

Cell and molecular biologists have proven that life is not the product of a random biological broth inside cells or of some mystical and amorphous stuff called protoplasm. They have, in fact, dispensed with the term "protoplasm" because it no longer has any useful meaning. To say that living cells are made of protoplasm is about as helpful as saying that an automobile is made of machinery. Understanding can come only from a more detailed view of what is happening in among the many components. Within just the last few decades the old views have been dramatically overturned. It is now clear that the property of life emerges from the cell for two utterly nonrandom reasons, two properties that keep the chemical and physical processes in every cell highly organized, which is to say two properties that keep alive the cells of all organisms more advanced than bacteria. Indeed, if either of these properties is lost, the cell dies, and if many cells die, the organism becomes diseased and, eventually, the organism as a whole dies. It is the integrity of these two properties that keeps us alive. (Bacteria, the most primitive known life form, lack these two properties and, perhaps for that reason, have not evolved into organisms of more than one cell.)

The first property is that cells have internal compartments, the organelles, with different sets of chemicals in different kinds of compartments. In all human cells (and indeed in all cells of organisms more advanced than bacteria), the compartments are enclosed by membranes—thin, two-layered sheets made predominantly of special fatty molecules. The membranes are like the glass of beakers and test tubes in a chemistry lab. But unlike glassware, most membranes completely seal the volume within them, allowing passage in or out only to certain specific molecules and then usually only the molecules that pass inspection by gatekeeper molecules embedded in the membrane. The gatekeepers—called receptors or, sometimes, docking proteins—peer out from the membrane at the passing scene, waiting for just the right molecule to come along. Then the receptor grabs the molecule. Or you can think of the receptor as a docking site shaped so that only one particular kind of molecule can come along and fit into the receptor. Once the two molecules embrace, drawn together by physical forces between one another, the receptor changes its own shape. The part of it that sticks inside the organelle then triggers some specific process. In some receptors, this process leads to the arriving molecule being taken inside the organelle, or even into the cell as a whole because similar receptors exist on the outer surfaces of cells. In other cases, the molecule stays outside but simply triggers some process within the organelle or cell.

The concept of receptors—first advanced in 1900 in one of the most brilliant speculations in science by the German biologist Paul Ehrlich—is one of the most fundamental in cell biology. Over and over again, biologists are finding out that many processes in life are the result of receptor molecules binding only to specific other molecules and then performing some action with them or upon them. Receptors are often thought of as locks that can be opened only by the right keys (Figure 2.2). If parts of the two molecules fit one another, they bind because of chemical attractions that depend on atomic forces that extend only a very short distance out from the surface of each atom making up the molecule. Because the forces are weak, the strength of the binding depends on the closeness of the fit and the size of the area where the two molecules come together. For example, imagine a customer trying out chairs in a furniture store. The more closely the chair (receptor) matches the shape of the customer's body, the tighter the binding. A customer trying out a stool will find it conforms to only a small part of his or her body. A curved chair with a back and sides, however, will produce a better fit, a better binding. Some chairs fit some customers perfectly and others poorly; in molecular biology the better-fitting chairs would be said to have high specificity.

Organelles also send out substances to be taken up by some other organelle in the cell. Most of this shipping is containerized; that is, the molecules are transported inside tiny bubbles of membrane, called *vesicles*. This process, too, is mediated by receptors. Each vesicle bears on its outer surface a receptor shaped to fit a complementary receptor on the organelle that is to receive the cargo. When the receptors lock, the vesicle fuses with the receiving organelle and the cargo is automatically dumped into it.

This way each organelle takes in only the substances it wants and then hosts specific chemical reactions involving the imported molecules. Then the products of those reactions, certain newly made substances, are exported from the organelle. Most of the larger bubbles and blobs that jiggle and lurch under Goldman's video-enhanced microscope are organelles and vesicles shuttling among organelles.

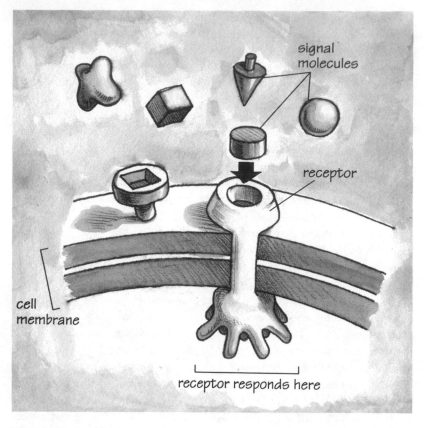

**Figure 2.2** Among the most important events in cellular life is the docking of a signal molecule with a receptor of complementary shape. Receptors stud the outer membranes of cells and the membranes of many internal organelles. When a signal molecule of the proper shape binds to the docking site, the part of the receptor inside the cell responds, typically by changing its own shape to activate some process inside the cell.

The second form of organization essential to life above the bacterial level of advancement is a transportation network that links the organelles, ferrying the vesicles from one compartment to another so that they can undergo a series of necessary chemical processing steps. This facet of intracellular organization is one that has come to light only since the 1980s. The system consists of an elaborate system of tracks that run in many directions within cells—long, filamentous highways that lace the cell like a three-dimensional subway system with thousands of lines. And, amazingly enough, cells also have molecule-sized motors that haul the cargo along the tracks. At any moment, thousands of chemically laden vesicles are being transported by tiny molecular motors from place to place in the cell.

Some researchers suspect that several major maladies such as Alzheimer's disease and Lou Gehrig's disease, formally known as ALS (amyotrophic lateral sclerosis), are caused by defects in the transportation system. An understanding of it could show the way to better treatments or preventives.

The cargo being shuttled about includes molecules manufactured deep within the cell that must travel to the faraway outer regions of the cell where they are needed. Inside other vesicles are molecules that the cell takes in on its surface—for example, nutrients and hormones received from the bloodstream—that must go to the cell's interior.

The longest highways of the sub-Lilliputian freight line are bundles of filaments inside the single-celled nerve fibers that run from the base of the spinal cord to the toes, a trip of three to four feet in tall adults. The molecular motors can make it in three or four days. The shortest hauls, lasting only seconds, are within almost all other kinds of cells, which have more compact shapes.

The discovery of how these molecular motors work—the discovery, in other words, of a fundamental mechanism by which life generates its own motion—was the product not of a sudden breakthrough by a lone genius but of many people working in groups over many years on problems in which there is a long historical interest.

Biologists had long known there was motion within cells. Many, studying cells under the microscope, often saw objects within that seemed to flow as if being swept by a current. Long ago biologists used to watch this streaming motion in their microscopes and think it one of the magical properties of life. It led them to imagine that the jellylike stuff inside cells, which they called protoplasm, was endowed with supernatural powers. Thomas Henry Huxley, Darwin's great advocate in the 1800s, used to give public lectures entitled "Protoplasm, the Stuff of Life." Even Gilbert and Sullivan helped make protoplasm famous. In their operetta *The Mikado,* they refer to "protoplasmic globules." Among the magical powers of protoplasm was its ability to host the flow of particles (they were not yet known to be containers), propelled by the intangible "vital force."

But science continued its long, slow turn away from the vital force theory, as laboratory experiments shed more and more light on entirely natural processes that could account for the wondrous properties of protoplasm.

One great leader in this movement was Albert Szent-Györgyi, a Hungarian immigrant to the United States who won the Nobel Prize for discovering vitamin C. The most thrilling moment of his long scientific career, Szent-Györgyi once said, was not winning a Nobel but seeing a nonliving laboratory preparation produce autonomous motion. It was

1934 and the scientist was working with some long, gelatinous fibers from a rabbit muscle. The fibers were known to be made of the proteins actin and myosin, which are primary components of muscle but in this experiment were little more than threads, the kind of stringy stuff a person might pull out of a chicken leg from the butcher. Szent-Györgyi also had a beaker of an unknown mixture of chemicals that had been extracted from mashed-up rabbit muscle cells. The minced muscle had been mixed with hot water and then the solid gunk was filtered out to leave in solution only those substances that would dissolve in the hot water, much as one would extract the flavor molecules from tea leaves.

Then Szent-Györgyi dipped the muscle threads in the extract. They quickly began to shrink, growing shorter and thicker. "Seeing motion, this age-old sign of life reproduced in vitro [in glass, literally] for the first time," he wrote many years later, "was the most exciting moment of my long scientific life."

Szent-Györgyi didn't know it at the time, but the hot-water extract contained one of the most crucial kinds of molecules in cells, adenosine triphosphate, or ATP. It is a molecule manufactured in all cells that, somewhat like a battery, stores and releases energy to power most of life's activities. (In Chapter 4 we'll see how cells make ATP from the energy contained in food.) The disembodied strings of muscle contained the molecular machinery to contract and did so as soon as they had access to energy to power the process. (In Chapter 9 we'll look in greater detail at how muscle works.)

Another historical path that led to the discovery of transportation systems in cells was the knowledge that things somehow flowed inside the long axons of nerve cells. For many years biologists had studied the unusually thick axon of the squid. Not only was it unusually large and easy to manage in the laboratory, it would live a long time after removal from its host animal. Researchers knew that if they injected special radio-active tracer materials into the axon at one end, the material would eventually show up at the other end. They also knew that the transport depended on the presence inside the axon of long, thin filaments called microtubules.

Microtubules form a vast meshwork concentrated near the nucleus and radiating out to all nether parts of the cell (Figure 2.3). Microtubules are made up of countless smaller subunits that, under the right conditions, spontaneously assemble themselves into tiny tubes. Microtubules were found to be essential to transport within axons because of experiments showing that if you treated a nerve cell with a substance known to prevent the subunits from assembling into microtubules, the radioactive materials would stay right where they were injected. But these experiments could

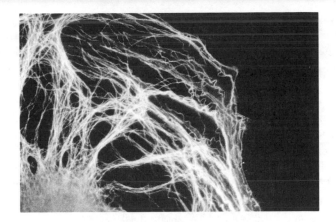

Figure 2.3a

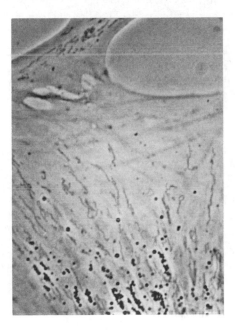

Figure 2.3b

**Figure 2.3** (a) The elaborate transportation network in a single cell. The picture was made by infusing the cell with molecules that stick only to tubulin, the subunit of which microtubules are made. Because those molecules glow when illuminated with one particular wavelength of light, they can be photographed. Everything else in the cell stays dark. The large fuzzy area is a dense network of microtubules enveloping the nucleus. (b) A cell in ordinary light. Microtubules are too thin to show up but some of the things transported along them are visible, including sausage-shaped mitochondria.

not explain how the motion happened. Some mechanical force had to be moving materials along the axon, but nobody knew what it was.

The answer began to emerge in the 1980s from rival teams of scientists at the Marine Biological Laboratory, the same place on Cape Cod where Jacques Loeb artificially fertilized sea urchin eggs earlier in the century. MBL's seaside location in the small village of Woods Hole and its access to an abundance of marine organisms for study, including squids by the thousands, have made it a favorite with biologists who have flocked there each summer for more than a century. One of MBL's buildings is named for Loeb. Albert Szent-Györgyi also spent much of his career at MBL. In fact, 33 Nobel laureates have worked at MBL, either as summer investigators or as students in the intensive summer courses the lab gives to many of the best and brightest in basic biomedical science.

Deciphering the transport system within axons would eventually open the door to a whole new understanding of how motion occurs, not just in squid nerves but in all cells of all species. It began with a serendipitous discovery in 1981 by the late Robert D. Allen, who came to the MBL every summer from Dartmouth College in New Hampshire. For years Allen had been studying the movement of vesicles down the length of nerve cells removed from squid. The squid's giant axon is a cylinder several inches long and a millimeter thick—about as thick as this o. He discovered that the vesicles moved in regular paths, along what he called "linear elements."

In the summer of 1981 Allen was teaching a course at the MBL and wanted the whole class to watch this motion simultaneously. To do this he aimed a special television camera into the microscope so that the image could be shown on a monitor. The system was an early forerunner of Goldman's now-standard setup.

Allen fiddled with the contrast and brightness controls on the monitor and was startled to see images that nobody had ever seen by looking directly into the microscope without the video attachments. The motion suddenly became dramatically clear. Little round objects could be seen shuttling in both directions along filaments running the length of the axon. It turned out that the video system was able to enhance the microscope image, magnifying and improving the contrast so much that it could reveal objects ordinarily too small to see with the microscope alone. Allen had found a way to see objects only a millionth of an inch across, one-tenth the size usually seen with a conventional microscope. Although electron microscopes can easily see such small objects, living specimens must be killed and dried first, eliminating any prospect of seeing the bustling activity of life. The video-enhanced light microscope made a squid's giant axon look like an aerial traffic reporter's view of a long, multilaned highway. Vesicles cruised in both directions. They moved not within the microtubule but along its outer surface.

Allen's observation of the moving vesicles yielded dramatic confirmation not just of motion within cells but of organized movement along fixed paths. Soon Allen found that an intact living cell wasn't even necessary. He and colleagues found the motion would still happen if he chopped the axon out of a nerve cell and squeezed its contents out, like a sausage from its casing, onto a glass microscope slide. The vesicles kept right on cruising along in this little dollop of gelatinous stuff as long as researchers added a bit of liquid containing ATP, the same energy source into which Szent-Györgyi had dipped the muscle fiber.

In such preparations of cytoplasm—a vague term that refers to the contents of a cell other than its nucleus—the microtubules, once neatly parallel in the axon, spill out into a spaghetti tangle and guide the course of vesicles in loops and bends. Where microtubules happen to cross, they sometimes form an interchange where vesicles may exit to the right or left. From the air, this looks not like one simple highway but the whole Los Angeles freeway system.

Over the next few summers MBL would be the scene of an increasingly exciting and sometimes antagonistic race among rival research groups to explain the detailed mechanism that made the vesicles move.

The answer came four years later when Tom Reese, an MBL regular from the National Institutes of Health in Bethesda, Maryland, and Michael P. Sheetz, then at the University of Connecticut, found that the job of transporting the vesicles—not only in nerve cells but probably in all cells—is performed by a special kind of protein molecule, which they named kinesin from the Greek word *kinesis*, meaning motion. Kinesin not only makes other things move, it does so by undergoing motion of its own—flexing and swiveling at points within its structure. The idea that molecules can change shape may seem odd to those who imagine them to be like rigid, stick-and-ball models. Some molecules are indeed as rigid as the models imply. But many, perhaps most of those that give life its distinctive properties, are flexible or, at least, as changeable as the shape of, say, a pair of pliers or a can opener or, for that matter, almost any windup toy. Like most machines, these molecules have rigid parts linked at flexible hinges. And, like everyday machines, the parts of these molecules can move with respect to one another only if an energy source is available. For simple machines, the energy source may be the muscles of the human hand. For workhorse proteins, the energy comes from ATP.

Kinesin is a motor. One end of the motor grabs onto the vesicle and the other latches onto the microtubule. Then the kinesin flexes and pulls the vesicle forward along the microtubule. The image is something like that of a man sitting in a rowboat and pulling it along a pier by using his arms and working hand over hand. The pier is the microtubule, the track of the transportation system. The boat is the vesicle, carrying a load of molecules to some other part of the cell. And the man is a kinesin molecule, a motor molecule. As the man flexes and swivels, the boat is hauled the length of the pier. Kinesin may also be envisioned as an inchworm, carring cargo on its back (Figure 2.4).

Sheetz, Reese, and company established that kinesin moves vesicles at a speed of about 15 inches per day. That's a glacial pace on the human scale of things, but it looks positively speedy in the video images of a single cell magnified ten thousand times. In a cell of typical size, it takes only seconds for kinesin to move objects as far as they need to go. Only in the very much longer axons does the trip take days.

As it happens, kinesin moves freight only in the outbound direction—from the nerve cell's main body through the axon to a distant structure called the synapse, which is where one nerve cell relays its signal to another. (Inside the vesicles are signal molecules, called neurotransmitters, that one nerve will release to act upon another.) Different vesicles, however, also move inbound along the same microtubules, propelled by different motor molecules. If kinesin is the "downtown train," some other motor must act as the "uptown train." That motor was discovered by yet another group that summered in Woods Hole, led by Richard B. Vallee of the Worcester Foundation for Experimental Biology in Shrewsbury, Massachusetts, an institution best known for developing the birth control pill in the 1950s. It turned out to be a molecule already known from another context and called dynein, from the Greek *dynamis,* meaning "power."

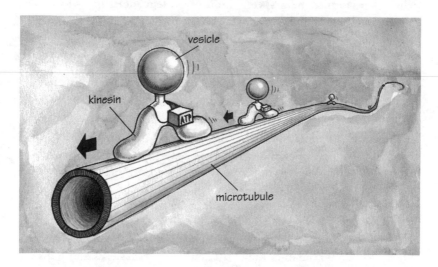

**Figure 2.4** "Motor molecules," such as the kinesin fancifully depicted here as an inchworm, transport parcels of cargo inside membrane-bound vesicles throughout the cell. Vesicles have receptors that bind to kinesin, allowing them to be carried as the motor molecule flexes to "walk" along the microtubule. ATP, the universal energy carrier within cells, supplies the force to make kinesin move.

Dynein had been known for years for creating a special kind of motion inside the hairlike projections, called cilia and flagella, that many cells possess. The inside of the windpipe, for example, is carpeted with cilia that beat in rhythmic waves. If dust particles are inhaled and stick to the mucus lining the windpipe, the cells of the lining protect the lungs by beating their cilia in coordinated waves that sweep the particles back up toward the mouth, where they can be swallowed harmlessly. The similar but longer structures called flagella are the propellers of some single-celled organisms and, of course, of sperm. The motion happens because inside each cilium and flagellum are parallel bundles of microtubules. Linking adjacent pairs of microtubule bundles are molecules of dynein. Like kinesin "walking" down a microtubule with a vesicle, dynein flexes and swivels to make one microtubule slide past another. The sliding motion makes the microtubule bundle bend, and since it is the skeleton of the cilium, the cilium bends. To imagine how this can work, think of a person standing next to a rubber flagpole and holding it with arms straight ahead. Both the person and the rubber pole represent microtubules. The person's arms are dynein molecules. Now imagine that the person reaches higher and higher on the rubber pole and pulls. The pole will bend and, assuming he or she is limber enough, so will the person.

Vallee and his colleagues found that dynein molecules are present not just in cilia and flagella, but also in the main body of the cell. When he added dynein to batches of microtubules squeezed out of squid giant axons, he found that the vesicles moved in the direction opposite to the way kinesin propelled them.

In 1990 Vallee found a third motor molecule, a protein he named dynamin, yet another variant on the Greek. In subsequent months and years still other scientists found still more different kinds of motor molecules. It is now clear that there are numerous molecular motors. Each is specialized to produce some kind of living motion—moving a particular kind of cargo over a particular part of the cell's internal transportation system or producing the several kinds of motion needed to carry out cell division. We'll deal with that phenomenon, one of the most dramatic and surely the most central in life's dynamic repertoire, in Chapter 6.

Since the discovery of kinesin, cell biologists have gained a deeper understanding of how motor molecules work. In the case of kinesin, for example, researchers have now seen individual molecules of it under the electron microscope. Its structure is superficially simple. At one end it has two more-or-less globular parts that contain the binding sites for microtubules. At the other end it has two larger structures that bind to the cargo. Linking the two ends is a rod with a hinge in the middle. At the microtubule-binding end are also shapes that interact with ATP, the energy source. The interaction liberates some of the energy stored in the

ATP molecule and uses it to make the same end of the molecule flex and swivel so that it can "walk" down the length of the microtubule.

The work on kinesin has even gone so far as to measure how strong it is, how much force a single kinesin molecule can exert when it is hauling a load. This is done with a device that uses the quantum mechanical forces generated by a narrow beam of laser light to grab a single microscopic latex bead and hold it still while one kinesin molecule tries to pull it loose. According to one estimate, a single kinesin molecule can pull with a force of five to seven piconewtons. Very roughly, that is enough force to haul two one-thousandths of a rice grain one-half an inch in one second. This is surprising strength in such a small machine. Suffice it to say that given the small distances and light weights inside a cell, a single kinesin molecule can haul quite a sizable load.

One of the interesting corollaries of the motor molecule discoveries is that these little machines can also read direction signs. Since one goes north, the other goes south, these mere molecules must somehow "know" which way to travel on a microtubule. The exact details have not been worked out, but it is known that microtubules have a built-in directionality—a head end and a tail end. Their subunits, like Lego bricks, go together in a way that produces the molecular equivalent of a direction sign that points one way—just as a Lego tower can be built only if the new bricks are oriented in the correct direction. The motor molecules somehow read this arrow and either move with it or against it.

Another of the mysteries confronting researchers is how a vesicle signals which way it "wants" to travel. Outbound vesicles must somehow allow themselves to bind to kinesin (the outbound motor), but other vesicles that are supposed to travel inbound must avoid kinesin and instead make use of dynein. Simple logic says vesicles must have a way to tell the difference. If they didn't, a vesicle could find itself the center of a tug-of-war between kinesin and dynein.

Perhaps, some researchers suggest, outbound vesicles have receptors, or docking proteins, studding their outer surface that are specific only for kinesin. And inbound vesicles have docking proteins for dynein. But, the scientists wonder, where do the docking proteins come from and how does the vesicle know which ones to use? As in so much of cell and molecular biology, it is possible to imagine how such a system might work, just as one might try to imagine how the insides of a car's engine are linked. One likely possibility, according to recent research, is that the receptor is part of the molecule being transported. If the cargo protein had the right chemistry, one end of it would stick out of the vesicle's membrane and could serve as a docking protein for a receptor that exists only on the vesicle to which it "wants" to go. This is the molecular equivalent of an address label.

One curious early finding was that kinesin's freight-moving ability is not confined to vesicles, at least not in laboratory experiments. It also will carry microscopic latex beads sprinkled onto the preparation. Kinesin will even try to move the glass slide, although the result is that the much heavier slide stays put and the microtubules slither across the glass.

Since the discovery of kinesin, work on molecular motors has expanded greatly, as have the discoveries. About ten different kinds of motors were known as of late 1995, with the number growing steadily. One of the field's current leaders, James A. Spudich of Stanford University, estimates that there may be fifty or more different molecules that function as motors on microtubules or on another kind of filament called actin. (Actin is best known for its role in making muscles contract, but also plays a key role in the ability of individual cells to crawl like amoebas. Chapter 3 will examine this remarkable phenomenon in detail.)

And, Spudich says, given the fact that cells host still other events that require motion (the movement along a DNA strand of an enzyme that reads the gene, for example), there may be as many as a hundred different kinds of molecular motor.

What all this means is that cells have the biological equivalent of a containerized cargo transportation system. Each vesicle is a container holding certain substances. The truckers are molecules that pick up a loose container and move it in a certain direction along the nearest track, a part of the microtubule filament network. Most vesicles are thought to carry the chemical equivalent of address labels on their outer surfaces. Vesicles appear to be hauled along the filaments until they encounter an organelle whose gatekeeper molecules—still other receptors—recognize the address label and take it off the track.

As some cell biologists imagine it, the process probably works like an imaginary postal system in which letters have addresses but are dispatched randomly in trucks and planes. At every post office, somebody checks the address. If the parcel happens to be at its destination, it is accepted. If not, it is tossed back and sent elsewhere.

Once the vesicle arrives at its intended destination, its membrane fuses with that of the organelle and the cargo is automatically dumped inside (Figure 2.5). Vesicle fusion, of course, must itself be a strictly controlled process, for if it were not, vesicles throughout the cell would fuse into one big vesicle, destroying the orderliness of the compartmentalization.

The discovery of the intracellular transportation system is significant, for it is only the third form of living motion that biologists have begun to be able to explain. The two previously known forms are the molecular interactions that make muscles move—the motion that swings a baseball

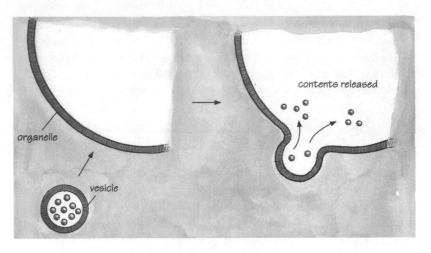

organelle

vesicle

contents released

**Figure 2.5** When a vesicle reaches its destination, receptors on its surface lock with others on the receiving organelle. This triggers the two membranes to fuse, an act that releases the vesicle's contents into the organelle.

bat or, for that matter, turns this page—and the interactions that beat, or wave, cilia and flagella. Researchers figured out many years ago the general outline of how a muscle works, and we'll deal with it in Chapter 9. More recently, cell biologists have begun to understand the beating of cilia and flagella—the rubber flagpole model. Now the new motor molecules provide the first evidence of how objects are transported within cells. All three systems, though they vary enormously in the scale of their results, work on fundamentally the same principle of small flexing molecules grabbing and pulling on long filamentous molecules.

The similarity suggests that all three may be the evolutionary descendants of one original method of generating living motion. In the primordial days when only single-celled organisms inhabited the earth, there may only have been the motor molecules that operated deep within cells. When cells evolved cilia and flagella, it may have been as a result of some modification of the motor molecules to operate these new structures. And, eventually, when multicelled organisms arose, yet another modification may have made muscles possible. Conversely, it is possible that each of these motors evolved independently but under such similar environmental demands that they converged on a resemblance that is only superficial, like sharks and dolphins.

In either case, the discovery of the motors may be historic. After all, it is autonomous motion—movement that is the result of some power within the moving object (and not simply the result of gravity)—that signifies "life" most dramatically. It is motion that "animates," providing the very root of the word "animal." In early science, the motive power was the "vital force." In its Judeo-Christian form, that force came from God, acting through the soul. When all detectable motion had ceased in a human body (the beating heart and the breathing chest were the traditional last signs), one could be certain that the soul had departed and that death was at hand.

If the ability of living things to move autonomously is one of the fundamental attributes of life—one of the features that distinguishes the living from the nonliving (and centuries of scholars have said that it is)—then one of the great mysteries of life is approaching solution. We know, almost down to the atom, how a muscle moves. We know generally what makes the cilia beat. And now we are beginning to learn what transports cargo-laden organelles inside every living cell. No doubt the remaining forms of living motion—the one that makes individual cells crawl about, for example—will be understood before long. When that happens, cell and molecular biologists will have made another major step in the philosophical revolution that is showing life to be a consummately natural phenomenon, one that the human mind can someday understand.

CHAPTER **3**

Scene One: The great, hulking predator brooded silently in the dim light, waiting for some signal that prey had blundered into its vicinity. The beast had roamed far in recent days but its hunts had failed. It had been many days since the creature fed.

Now it seemed unable to decide where to go next. Such an amorphous blob of an organism—it has no discernable brain—could not be "thinking" what to do, but it definitely looked indecisive. First it would push a flattish, gelatinous tongue out from its central mass, as if to test the possibility of heading in one direction. But then the protrusion would shrink back, disappearing like a wrinkle on a pulled bed sheet, and another lobe of jelly would extrude itself in a different direction. But it, too, would fall back.

The predator would not move. Not because it had no arms or legs as we understand them in the giant, human-scaled world. The predator was a cell, isolated from some person's body, and it had other means of locomotion. It could quickly assemble within itself the functional equivalent of bones and muscles any time it "wanted." This creature was a *macrophage,* one of the free-roaming types of white (actually colorless) blood cells that functions as a part of the immune system and one of the most fascinating types of cells in the human body. Macrophages inhabit the bloodstream but can also slip through tiny pores in the walls of blood vessels and wander in other tissues of the body. Macrophage is Latin for "big eater." As it happens, macrophages maintain the ancient ways of their protozoan ancestors by creeping about their habitat and feeding on other organisms. In the human body macrophages swallow invading bacteria and viruses whole the way an amoeba does, by oozing around them and engulfing them. Macrophages are also the body's garbage collectors, and sometimes even its undertakers, eating aging or fatally damaged cells of other types. One big job is to dispose of decrepit red blood cells. Macrophages in a typical human body collectively put away an estimated 100 billion red blood cells a day.

When a few bacteria were placed in the beast's chamber, which was being observed through a microscope, the indecisive cell quickly made up its mind. Somehow it sensed the presence of food. The cell pushed a pouch of its membrane toward the microbial morsels and began gliding across the chamber's clear plastic floor.

Macrophages are basically amoeba-shaped but their outer membrane is wrinkled and folded into hundreds of tiny lumps and ruffles. When a macrophage travels, it extends some of the ruffles closest to the floor, pushing them out like a huge, lumpy foot. Cell biologists call the lobe a *pseudopod,* or "false foot," a term first applied to the lobes of amoebas, which move much the same way. The underside of the lobe is studded with protein molecules that grab the surface below. The macrophage appears to pull itself along, the central mass of the cell riding behind the false-footed lobe like the shell of a snail.

When the tiny beast in the chamber reached one of the unwary bacteria, its pseudopod immediately changed form, flowing around and over its sausage-shaped prey. When the main body of the cell touched the bacterium, it too began to change shape, embracing the microbe in a death grip that would soon engulf it and swallow it whole. The microbe put up no struggle. Now the cell rested, apparently satisfied.

Scene Two: The two-day-old embryo of a chicken, dissected out of a fertile egg, lies in a tiny well in a plastic laboratory dish under the lens of a microscope. The embryo, quite unaware that it is no longer under the protection of a shell, continues to carry out its program of constructing a chick. Right now, however, the future chicken is hardly more than a simple tube made of many cells, no bigger than a hair from a one-day beard. This so-called neural tube will be the foundation of the architectural hallmark of vertebrates—a spinal cord enclosed by a spinal column of bony vertebrae. The neural tube is one of the earliest structures to form not only in the chicken embryo but in the embryos of all vertebrates, humans included.

At first all that can be seen is the neural tube. But over the next few hours a dramatic change appears. The tube is still there, but now hundreds of free cells are racing away from it. Once perched on the tube, they have now climbed off and are crawling across the plastic, striking out for what would seem to be faraway destinations at all points of the compass. The image is like a bird's-eye view of passengers scurrying away from all doors of a stopped train.

Embryologists call these disembarked passengers *neural crest cells.* Early in embryonic development they emerge from the top or dorsal side of the neural tube and glide away, heading off to pursue a bewildering variety of fates in what will become other parts of the embryo. Many of the wandering cells will develop into various parts of the nervous system; others will produce cartilage and bones of the skull. And some will even take up residence in the skin and become the cells that make melanin, the pigment that gives skin its color.

In the plastic dish, cut off from its supply of nutrients in the yolk, however, the embryo cannot live long enough for most of this to happen. But the cells nearest the tube, within a matter of hours, bunch together in characteristic patterns to form rudimentary components of the nervous system called *ganglia*.

Scene Three: Another dish of cultured cells—this time human skin cells—rests on the stage of a microscope with a video camera peering through the lenses. The picture on a monitor shows what looks like about a dozen fried eggs. The gray lump in the middle, where the yolk should be, is the nucleus, and around it is a border, nearly transparent and seemingly stretched out tight over the plastic surface, stretched far in some places, hardly at all in others. The cells appear to stand still, as immobile as, well, fried eggs stuck to the frying pan. If these cells move, they are not speed demons.

The video camera is wired to a computer and a video-disc recorder, a device that encodes pictures digitally and stores them on a compact disc just as a CD stores sounds. Every sixty seconds the camera takes a picture and stores the image in the recorder. After recording all night, the still images are played back much faster than they were recorded. Ten hours is compressed into half a minute. This is the same kind of time-lapse photography that makes a flower bloom in five seconds.

Now there is no mistaking the power of life. The fried eggs do move. Part of the egg white, the thin border, lifts up off the bottom of the dish and waves in the air or, rather, in the watery nutrient medium. It ruffles its edges as if testing the space ahead, and the cell creeps in that direction, gliding and ruffling. Some cells look decisive. They fairly zoom across the video screen, their ruffles undulating ahead of them. The hind ends of the moving cells have long tails; spots of the cells' undersides stayed stuck to the bottom while the rest of the cell moved on, stretching the gelatinous blob of the cell into a long filament. Eventually, the adhesion breaks, and the tail is quickly pulled into the main cell body. Sometimes, however, the adhesion is too strong, and the power of the cell's forward motion rips a bit of the tail off. The cell's membrane instantly reseals the wounds, leaving a tiny blob of living tissue behind. Some cells are not so determined. They dither. They stretch out their borders in one direction and ruffle a bit only to pull back and reach out in some other direction (Figure 3.1).

Sometimes vigorous movers collide. They seem to cringe briefly, pulling back their ruffles as if startled. Then the cells back away, ruffle their edges on a different side and glide apart, heading in new directions, often along diametrically opposite paths. Every now and then a scurrying cell stops, pulls in its ruffle and all the rest of its border. It lets go of its grips on

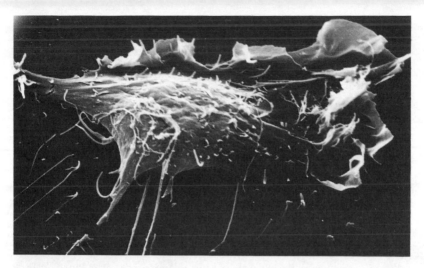

**Figure 3.1**  A kidney cell ruffles its leading edge as it crawls over the bottom of a culture dish. The ruffle's function has not been established, but it may be the cell's attempt to test the environment for chemical signals or surfaces on which to crawl. In contrast to the flat surface of a culture dish, cells in the body are surrounded by cells or other structures with which they normally interact. Filaments at the hind end of the cell are places where the cell did not let go of the surface as it moved ahead and the membrane was stretched. (Courtesy of Guenter Albrecht-Buehler)

the dish and rounds up into a ball. There is some jostling inside the ball, and then it bulges in opposite directions and, blip, the cell pops into two smaller balls. The cell has divided. The two balls quickly spread, flattening back against the dish and move away from one another.

In all of these episodes, which took place in various cell biology laboratories, blobs of chemicals, called cells, have somehow been able to pick themselves up and travel under their own power to new locations. They were being studied for answers to one of the most challenging questions in modern biology. How does life produce the motion that animates it? How does a single cell "decide" where to go? How does it generate the force that propels it? The vitalists of old were content to ascribe the power of locomotion to a supernatural agent, but biologists today know that motion is natural, the product of physical, chemical, and mechanical forces inherent in the structures of specific molecules organized in specific ways.

In the mid-1980s cell biologists saw motor molecules such as kinesin pick up objects and carry them about, not just in living cells but even after the individual molecules—naked and nonliving—were removed from cells and placed on a microscope slide. There was no "life" in any conventional sense in the smear of chemicals on the slide, and yet assemblages of those molecules did move. The animating force was somehow present in mere molecules.

And many years earlier, biologists saw how muscle cells produce their motion. That work began with experiments like Albert Szent-Györgyi's half a century ago, when he dipped some isolated threads removed from muscle fibers into a solution that contained the energy source ATP and saw the threads contract. That motion, scientists would eventually establish, was a result of one kind of protein called *myosin* acting like kinesin, grabbing onto another kind of protein filament and pulling. Kinesin pulls specifically on microtubules. Myosin pulls on a different kind of filament called actin. It is now well established that in muscle cells both actin molecules and myosin molecules are braided into separate filaments that lie parallel to one another. Myosin can pull because it has little projecting "heads" that act like kinesin, grabbing the actin and pulling on it. Biologists now understand most of the details of how the molecules of a muscle cell work together to produce their kind of living motion. (More on how muscles work in Chapter 9.)

"Muscle was worked out quite a while ago. So it is logical to wonder whether non-muscle cells move because they have the same stuff in them," says Edward Korn, scientific director of the National Heart, Lung and Blood Institute, which is one of the National Institutes of Health. Korn's lab specializes in the study of myosin in nonmuscle cells. "Well, they do. It looks like all cells can make both actin and myosin. And when we look at motile cells and ask where the actin is, we see they have it in the right places to do the work."

That place is a thick layer just inside the cell's outer membrane. There, cell biologists have found an elaborate meshwork of actin filaments that form part of the cell's skeleton, or *cytoskeleton* (from the Greek *kytos* for "hollow vessel" and now generally used to mean cell). (Actually, there are three diferent networks of filaments that are called cytoskeleton. One is the network of *microtubules* such as those that serve as tracks for kinesin motors. A second is the less well-understood network made up of what are called *intermediate filaments,* of which more later. The third is the *actin meshwork* that defines a cell's shape and enables it to move.) Unlike a bony skeleton, the actin cytoskeleton can change shape continually, which it must do if a mere blob of a cell is to change its shape continually, extending and retracting such special structures as *pseudopods* and *ruffles.* It is as if the cell could construct bones (actin) and muscles (myosin) on the spot, exactly when and where they are needed.

Lying just inside the cell membrane is a meshwork of short actin filaments, thought to be cross-linked in some way that makes them into a relatively firm girderwork. The meshwork is a living version of a building's space-frame—an intricate system of interlinked, triangular braces and trusses that define its shape. The cell's space frame appears not to be a geometrically perfect girderwork but an irregular meshwork that nonetheless has some rigidity. In the cell, this network of actin filaments holds the membrane and thus determines the cell's shape.

But unlike metal girders of fixed length, actin filaments can grow or shrink at any time because they are polymers—chains of actin subunits, or monomers. Most cells contain large amounts of actin, some as monomers and some as polymers of various lengths. Somehow the cell regulates just how much of its actin becomes polymerized. Part of the control is carried out by certain other molecules that bind to actin subunits, keeping them from linking up to one another or from joining existing polymers.

Many cultured cells that do not move tend to spread out flat against the bottoms of their culture dishes, assuming the fried egg appearance or sometimes looking more like a floppy starfish or even a crescent moon. As the cells spread, their actin polymerizes into very long filaments. Also a great many actin filaments lying in parallel become tightly bunched together, forming great bundles that reach from one edge of the spread cell to the opposite edge. The bundles are often called stress fibers. Cell biologists can often see these huge fibers in the microscope. They show up even more easily when the cells are stained with fluorescent chemicals that penetrate the cell and bind only to actin. Ultraviolet light is shined on the cells while they are under a microscope, and the stress fibers glow brightly. Scientists have also seen that at the ends of the stress fibers are anchoring points where the cell is stuck down tightly to the bottom of the dish, like a tent pegged to the ground. This anchoring is carried out by special proteins that link actin to other proteins that poke through the cell membrane and bind to the surface the cell lives on (Figure 3.2).

But when the time comes for the cell to divide—in many cells there is a biological clock that periodically tolls this moment—the cell releases much of its grip on the dish and the actin stress fibers break down. As they disassemble, the spread cell rounds up into a sphere, presumably pulled to that shape by the surface tension of the outer membrane. Sometimes the gripping molecules do not let go, and as the cell rounds up, patches of its membrane are stretched into long, thin tubes that look like spikes linking the now-spherical cell to its stubborn attachments to the surface. When cell division is complete, the daughter cells move apart and eventually each spreads out again. The two new cells reassemble their actin into long filaments. These become bundled into stress fibers that reestablish their grip on the bottom.

When the cell makes ready to travel, it also breaks down most of its heavy actin fibers and redeploys the actin into the dynamic space-frame version of its cytoskeleton. Some actin is probably also held in the monomer form as raw material for construction of new space-frame "bones" as needed. The gripper molecules may let go of the "ground" beneath the cell or they may actually stay stuck and be ripped out of the cell's body as

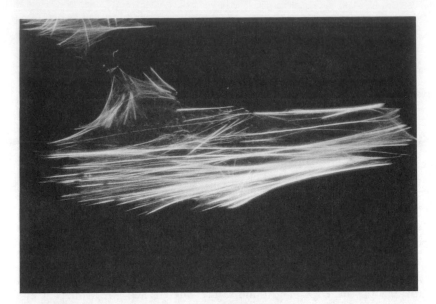

**Figure 3.2a**

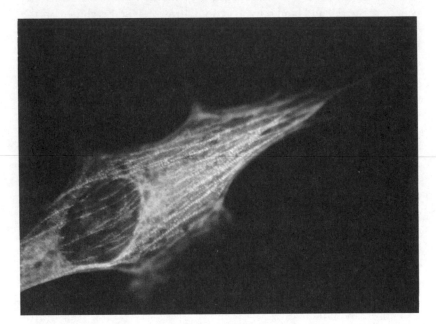

**Figure 3.2b**

**Figure 3.2**  (a) A cultured cell that has been treated with a chemical that binds only to actin. The chemical also glows when illuminated by a particular wavelength of light, in this case showing that the actin is organized into long, nearly parallel fibers. Many cell types organize their actin into this form when they are not moving. (b) A similar, nonmoving cell, this time stained with a chemical that binds only to the motor molecule called myosin. The image suggests that myosin occupies points along the same kind of actin fibers that can be seen in the other cell.

it moves on. Fortunately, because of their fluid and self-adhering nature, cell membranes instantly seal themselves to heal the momentary wound.

Bones alone cannot produce movement. There must be muscles too. Within the cell, myosin is the muscle. The exact mechanism that generates the cell's ability to move has not been fully worked out, but the idea is that the myosins act like motor molecules, pulling on the actin filaments—much as muscle pulls on bone—thus re-forming the cellular space frame into new shapes. Actin segments in the girderwork are not firmly riveted to one another but are linked by molecules of a short, rod-shaped protein that can bind actin at each end. The exact molecular process by which cells remodel their actin space frames is not well understood, but one possibility is that myosin binds to one actin filament and then "walks" up or down another, forcing the actin meshwork into new conformations. As the myosins pull on the actins, they could change the shape of the overall structure by first pushing the cell surface out and forming the pseudopods or ruffles typical of motile cells, and then by pulling the pseudopods back in.

Duke University's Michael Sheetz, one of the discoverers of the motor molecule *kinesin,* has investigated exactly what goes on in one specialized form of cell motility—the ability of a nerve cell to sprout a long arm, or tentacle, that reaches farther and farther into the space around the cell. Sheetz's model of cell motility relies heavily on myosin, or perhaps some yet-unknown motor molecule like it, to push and pull actin filaments inside cells. He derives much of his inspiration from time-lapse movies of cells exercising their motile powers.

"I can sit and watch these for a long time," Sheetz says as he gazes at a video monitor showing a cell moving about while a time-and-date digital counter clicks away in the lower left corner of the screen. "You begin to get a feeling, a sense of what might be going on. It's fascinating. This is behavior at a level we've never been able to study before."

The star of Sheetz's movie is a part of a nerve cell called a neural growth cone. Inside the human body, and indeed inside all multicelled animals above a certain tiny size, nerve cells are constantly sending out long, slender arms that snake off into the distance in search of some other cell. It is the job of a nerve cell, or *neuron,* to transmit a signal to another cell some distance away. The neuron finds the faraway cell by sprouting branches, called filopodia (for "filamentous feet") during embryonic growth to establish the body's wiring pattern. All through adult life neurons sprout filopodia to search out new linkups, which may be one way the brain and body learn or improve certain skills. There is evidence that if the linkups are used repeatedly, they become stronger and more linkups may form along the same path. If they are not used, however, the filopodia eventually shrink back, creating a physical change in the anatomy of a brain cell that may be at the root of the process of forgetting. Basically, the message is, "Use the neuronal link or lose it."

At the tip of the filopodium is the so-called growth cone, the active site of elongation. The neural growth cone demonstrates yet another example of how life can generate within itself the motive power, the animating force that enables the cell to extend its reach, and its biochemical influence, into new territory. It is one thing for cells to slither about like individual animals. It is yet another for neurons to keep their "cell body" in one place but to send off ever-lengthening, often-branching tentacles to make contact with other cells. This is the ability that turns a collection of neurons into a brain, each neuron in contact with thousands of others.

In Sheetz's movie a neural growth cone sprouts several new filopodia. They are straight, looking as if a rod were lengthening inside them, stretching their outer membrane ever farther. Then one sprouts a branch at almost a right angle to the main arm. The free ends of the filopodia wave ever so gently, but farther back, nearer the neuron's main "body," the filopodia are still.

Sheetz thinks actin filaments are polymerizing inside the filopodia, forming their skeleton. But he sees evidence of "muscle" too, probably a myosinlike molecule. Every so often along its length, a filopodium anchors itself to the surface below, forming some kind of structure that grips the bottom of the dish. Sometimes, however, a filopodium will— apparently mistakenly—anchor its tip. When that happens, the movies reveal an intriguing result. Very soon after, the ramrod-straight filopodium buckles. Between the two anchor points, it forms a bend that grows sharper with time. It is as if the filopodium is trying to grow longer but can't because it is pinned down at both ends.

"This doesn't prove anything," Sheetz cautions as he rewinds the tape and replays the sequence of a filopodium buckling. "But you look at this and you do get a strong sense of what might be happening."

What seems to be going on is that some force-generating mechanism within the filopodium has gone awry. It is supposed to extend the filopodium into new ground, but instead it is thwarted because the tip has become pinned down. The actin filaments keep growing longer, and the force-generator, or muscle, keeps pushing on the actin skeleton as if to make the filopodium longer, but the force is so great, it buckles the actin rods inside.

Sheetz envisions a mechanism that might work like this: Lining the inner face of the filopodium's outer membrane is a meshwork of some protein, not actin, as is the case with many other cells, but perhaps a molecule called spectrin, which forms a thin but relatively rigid meshwork lining the membranes of cells. The spectrin meshwork is thought to be anchored to the membrane at many points so that the flexible membrane is held in some particular shape that conforms to the shape of the spectrin meshwork. For example, the characteristic shape of the red blood cell—a disk with a very thin center—is determined by the spectrin meshwork.

Running down the middle of the filopodium is a filament, or cable, of actin. Linking the actin to the spectrin, Sheetz speculates, are lots of myosin molecules. The arrangement would be something like that of, say, a dozen men holding a long wooden pole, six on each side. The men, who stand still, are myosin molecules and their feet are anchored to the spectrin and the cell membrane. The pole is an actin cable. At a signal, all the men push the pole in the same direction. The pole moves, stretching the filopodium's membrane (which encloses the men and the pole) in the same direction. This is the way the filopodium normally moves. But now imagine that the forward tip of the pole is pegged to the ground and the men push in the same direction. If they push hard enough, it will buckle or break.

Sheetz emphasizes that his model is largely conjecture inspired by a movie. But it is a movie that shows living cells performing behaviors that can seldom be recognized in still images.

Although today's cell biologists rely heavily on video microscopy with elaborate electronic image processing to bring out details never before seen, the practice of making time-lapse movies of cells in motion goes back almost two generations, to a time when the 16-mm movie camera was the state of the art.

The alternative approach of relying on still images poses a problem that has been compared to that of a Martian trying to discover the rules of football by examining a few still photographs from different games. A movie of one whole game would be far more meaningful.

Along with a method of generating force, the laws of elementary physics dictate something else about cell locomotion. A cell could not move over a surface unless it had some form of traction, some way of gripping the surface so that it could use its motor molecules to push against it. Trying to move without a surface-gripping mechanism would be like trying to walk on wet ice. In the same way, the cell must be able to grip the bottom of its culture dish—or any surface within the body—with part of its anatomy while it remodels its actin cytoskeleton so as to extend a filopodium or pseudopod or ruffle.

Cells are not heavy enough to count on friction to grip their microscopic roads, but it is well established that they can grab certain kinds of surfaces around them. All cells possess special molecules that project through their outer membranes and which bind to one or more of the natural materials that line most surfaces in the body and fill intercellular spaces. In fact, many cells make this lining in the first place and release it as a secretion through their outer membranes. The lining, called *extra-cellular matrix,* is a combination of several special proteins (they go by such names as laminin, fibronectin, and one that may be better known to many as a supposed miracle ingredient in cosmetics and shampoos, collagen). These proteins are laid down on whatever surfaces are available outside the cell, often in highly ordered patterns. They stick to the surfaces and provide the cells with something to grip (Figure 3.3).

The grippers are proteins shaped specifically to bind to extracellular matrix proteins. These receptors are mounted in the cell's membrane, one end sticking out to meet the outside world and the other end sticking inside, where it is thought to be bound to the actin meshwork just inside and perhaps also to other cytoskeletal elements such as spectrin. Because the receptors are integral membrane proteins, they have been named integrins. There are many different kinds of integrins, each capable of gripping only one or a few matrix proteins.

Cell biologists generally believe that for a cell to move, it must do two things—stick to some fixed surface outside the cell and then use its internal motor molecules to remodel the actin cytoskeleton so as to push the leading edge of the cell forward, leveraging that forward motion against the gripped surface. The leading edge—such as the ruffle of the cultured skin cell in Scene Three—reaches over the surface without making contact until it has stretched ahead some distance. Then it touches down, and its integrin grippers grab the surface. The motor molecules push the ever-reshaping actin meshwork farther ahead. The receptors at the cell's hind end either release their grip or are physically cut off from the cell. Nobody knows how cells let go. Or, as was mentioned earlier, the grip does not release, and the cell develops a thin "tail" stretching out behind it. Some cells in culture rip out their grippers as they move, leaving behind long trails of them. This means the cells must make new integrin to maintain their normal complement.

One of the obvious, but profound, observations from many studies of cell motility is that cells usually move in a coordinated way. The macrophage, as in Scene One, does not wander aimlessly but deliberately seeks out and devours its prey. The cell in an embryo glides from one part of the nascent organism to the site of its future role. The skin cells healing a wound crawl out onto the exposed surface and array themselves in a new layer. In every case, the cell behaves in an organized way, responding appropriately to signals it receives from the surrounding environment. The actin–myosin mechanism appears to work in concert throughout the cell, propelling it in one direction. If there were not some sort of coordinating mechanism, nothing would keep each part of the cell from "deciding" to

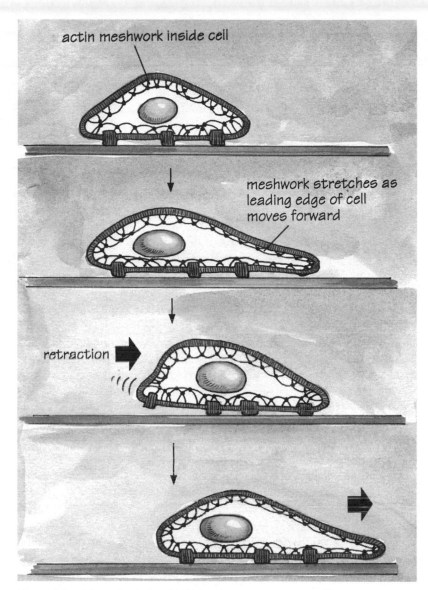

**Figure 3.3** A hypothetical and highly schematic suggestion of how the actin and myosin meshwork might function in a crawling cell. The precise architecture of the meshwork is completely unknown.

travel in its own preferred direction. Such a cell might rip itself into pieces, as it heads in more than one direction.

Such observations and conjectures are simple, but their implications are profound. Each cell can take in information about its circumstances and respond in a purposeful way, carrying out its crawling ability in a manner that seems coordinated throughout the cell.

"I say that this is a form of intelligence and that, therefore, the cell must have a control center, a brain," says Guenter Albrecht-Buehler of Northwestern University's medical school in Chicago. Albrecht-Buehler is a physicist who turned to cell biology some years ago, bringing an approach to the problems of life that most mainstream cell biologists consider "way out" but which some consider at least provocative and which a few regard as visionary.

"Most cell biologists only care about the structures they see within the cell—whatever they can see under their microscopes. But I say that's not what's important. What's important is the information being processed within the cell to produce coherent behavior on the part of the whole cell." The situation is something like trying to understand a computer by examining only its hardware. The shapes of computer chips and keyboards and disk drives reveal almost nothing about what the machine does. To understand that, you must know the information entered into the machine, how it is shuttled about inside and processed into some result.

Albrecht-Buehler has done many experiments on the locomotor behavior of cultured cells to learn how they take in information about their surroundings and respond to it. He finds that cells behave in ways that are uncannily like the behaviors that a whole animal would display. In one experiment, for example, Albrecht-Buehler put isolated mouse cells in the microscopic equivalent of a maze. The maze was a glass slide coated with a thin film of microscopic gold particles, much thinner than a cell. Albrecht-Buehler then swept out tiny gold-free paths, creating a pattern like the streets of a city laid out in a regular gridwork.

Cells dropped onto such a surface immediately show a predeliction for traveling on the gridwork of "streets" even though the completely inert and nontoxic gold particles are comparatively tiny, no bigger in relation to a cell than a piece of gravel is to a human. In fact, if they have no choice, cells readily crawl over surfaces completely covered with gold particles. Mouse cells will wander widely on an all-gold surface, engulfing the particles as they move and leaving cleared paths behind them. On the gridwork surface, however, the cells prefer to stay on the streets. They travel down one street in a straight line until they come to an intersection. Then they explore each option, probing tentatively into each of the three possible routes before making a decision and either turning a corner or continuing in a straight line.

"This is not the behavior of a mere automaton," Albrecht-Buehler says. "The cell is exercising a form of intelligence, testing each possibility and only then moving on. Somehow the cell is taking in information about its environment and it is processing that information. The cell does not try to go in all directions at the same time."

This is a key point, Albrecht-Buehler says. If you tear a cell into pieces, many pieces will continue behaving on their own for a while, crawling about as if they were tiny, whole cells. Each moving fragment can act with autonomy. But when the pieces were part of a whole cell, they were not autonomous; they were subject to some central controller—some mechanism that Albrecht-Buehler, to the dismay of his colleagues, calls the cell's brain.

The "brain," Albrecht-Buehler says, is not the cell's nucleus or the genes it contains. All biologists accept that genes are repositories of information that govern the cell's chemical and behavioral potentials, but that they are essentially passive players in the moment-to-moment behavior of the cell. Moreover, you can take the nucleus out of a cell, and the cell will continue to behave normally, responding appropriately to environmental cues. Eventually, as the cell uses up some of the proteins it had previously made and needs to consult the genes again for the blueprints on how to make new proteins, the cell without a nucleus will die. (We'll look at the nucleus in more detail in Chapter 5.)

Albrecht-Buehler speculates that the cell's brain, its behavioral controller and the seat of its intelligence, lies in a complex, poorly understood structure known as the centrosphere. In most cells the centrosphere, more commonly called the *centrosome* or cell center, is situated near the nucleus and is the focal point of the part of the cytoskeleton made of microtubules. From the centrosphere a dense network of microtubules reaches out—like a nervous system, Albrecht-Buehler likes to say—to all points on the periphery of the cell. The great mystery, if he is right, is the form of the information the cell might be taking in. What are the sensory organs of a cell? Does it have eyes and ears, or feelers? How does it transmit the signal, once received, to the centrosphere? How are the signals processed, and then, assuming a "decision" is made to respond, how is that decision carried out? How would a cell tell its actin-based cytoskeleton to remodel itself? Somehow, it seems obvious, the activity must be coordinated if the cell is to move in a coherent fashion.

Most cell biologists imagine that these problems are taken care of by the aggregate of thousands or millions of chemical reactions happening throughout the cell, each taking place more or less independently. Central coordination, they would say, is an illusion produced by the fact that as similar chemical conditions prevail throughout the cell, or throughout major regions of the cell, the chemical reactions that lead to behavior automatically proceed in the same way. The intelligence of the cell, most experts in modern cell biology would say, is distributed among all its molecules and is implicit entirely in their shapes and chemical affinities.

Even Albrecht-Buehler concedes that he cannot yet prove the cell has a central mechanism for control of its behavior. But he has found preliminary but tantalizing hints that a key role is played by two very strange-looking structures that lie within the centrosphere. These are the so-called *centrioles*—two cylinders, each looking as if it were made of lead pipes welded side by side. Each pipe is, in fact, a microtubule, the same structure that forms the highway used by kinesin motor molecules. In the centriole, however, life has rearranged microtubules into a very different form, one whose geometric precision looks anything but "organic." Two centrioles exist within the centrosphere of every animal cell that moves. For example, virtually all animal cells, except the ovum or egg and certain cells that lack nuclei, have centrioles. And all plant cells lack centrioles, except for a few unusual plant sperm cells, which produce centrioles when they must move. Many plant sperm cells have lashing tails, like human sperm, but some move by amoeboid motion, creeping toward their goals like an animal cell in a culture dish.

Each centriole is made of exactly twenty-seven short microtubule pipes welded into nine groups of three each. The nine triplets are arranged in parallel and linked along their lengths by other small molecules so as to form a hollow tube. In cross section, the triplets look something like the petals of a flower. Down the center of the cylinder are still more structures, too small to be seen clearly in the microscope, but there appear to be spokes of some sort linking the triplets to still other structures in the center of the cylinder. Also, the triplets are slightly twisted, making a side view of the centriole look a little like a barber pole. Moreover, every motile cell has two of these centrioles, and they are always arranged at perfect right angles to one another (Figure 3.4).

"This kind of universal design cannot be an accident of evolution," Albrecht-Buehler says. "It must have arisen to serve a purpose."

In one of the longest leaps of speculation in cell biology, Albrecht-Buehler suggests that the centrioles are the cell's eyes. He has calculated that the paired centrioles have exactly the right geometry to "see" point sources of electromagnetic radiation—such as the faint infrared signals (heat) emitted by another cell. The thin space between the triplets, the angle of the barber-pole twist, and the right-angle orientation of the pair all fit with the idea that centrioles are detectors, telling the cell where there may be other cells. It may, or may not, be relevant here that when cells are crawling, their centrioles are positioned on the front side of the nucleus, the side facing the direction of travel.

Albrecht-Buehler concedes he has no idea how the cell might then process this information, but he cites a common observation of cell biologists watching cells crawling about the bottom of a culture dish. If two

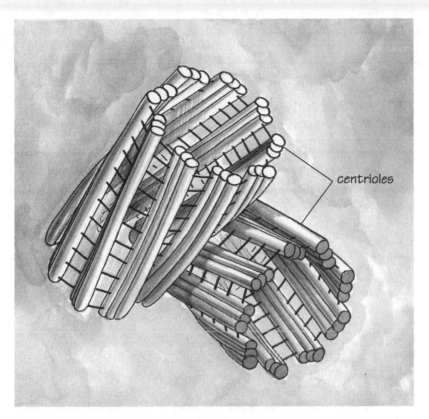

centrioles

**Figure 3.4** One of the most enigmatic of organelles is the centriole, which looks more like something made in a machine shop than in a living cell. It is made chiefly of short segments of microtubules that have become welded side by side in triplets and then twisted. Their best-known role is to help separate duplicate sets of chromosomes during mitosis.

cells come within a certain distance of one another, they often change course, each steering toward the other. Often when cells come within this range, they quickly extend a ruffle or pseudopod toward the other cell. Somehow, it seems, they can detect one another's presence. The cells meet, touch briefly, and then, most often, recoil and veer apart.

Centrioles and the microtubules that emanate from their vicinity have also been linked—in a more speculative way—to what is probably the most difficult phenomenon of life to explain—consciousness. In recent years, for example, such disparate scientists as Francis Crick, who helped discover the double helix shape of DNA, and Roger Penrose, a physicist who usually works on black holes and relativity theory, have suggested that these organelles may be the agents of consciousness. They offer no direct evidence but cite the coordinated behavior of cells as evidence that they have some kind of information-processing ability. They note the peculiar nature of the centrioles and even suggest that the microtubules that emanate from them could be roughly analogous to a nervous system within each brain cell. If the microtubule network did, indeed, function in this capacity, it would give brains a vastly larger order of complexity than is already evident in the organization of neurons.

Neither Crick nor Penrose purports to know how microtubules might generate consciousness, but they do cite an intriguing property of these long filaments that suggests they could act as carriers of information within cells, perhaps analogous to the wires of a computer. Microtubules, as we have seen before, are polymers made up of many subunits, or monomers, called *tubulin*. Each tubulin monomer can change between either of two slightly different shapes, depending on the behavior of one of its easily detached electrons. The propensity of a tubulin to exist in either shape is dictated by its neighbors. Thus, if one tubulin flexes to the other shape, its next-door neighbor will flex, too. As a result, the reversal of shape propagates down the length of the microtubule, causing the free electron to be transported from one end of the filament to the other. This is a form of electricity, which is more familiar as the flow of many electrons along a wire. With that slender analogy to wires and with the fact that the behavior of individual electrons is subject to the unpredictable influences of quantum mechanical phenomena, the scientists suggest these humble filaments may be part of the mechanistic basis of consciousness. No doubt it will be a very long time—if ever—before consciousness is explained. Until then, one of the more immediately approachable phenomena of life—how cells generate the coordinated mechanism that allows them to crawl—remains one of the hot areas of current cell biology.

The explicit reason is that motility plays a fundamental role in many key processes of human health and disease. During embryonic development cells crawl all over the embryo, gliding over, under, and around one another to reach destinations within the embryo where they will settle down and pursue their different fates in specialized tissues. During life several kinds of cells move about the body. There are the macrophages seeking not only bacterial food but consuming dead and decaying human cells. There are the cells that take up old bone and others that lay down new. There are neurons that send out projections to rewire the brain and nervous system as the body learns new skills. And cell motility is central to the problem of cancer. As long as tumor cells stay in one lump, growing right where their ancestor began multiplying out of control, the tumor is said to be benign. But one of the powers acquired by "malignant" cells is the ability to move. Such cancer cells do not just spread passively. They can also travel of their own primitive will. They acquire specific abilities to let go of their attachment of neighboring tumor cells and they crawl off to seed new tumors. If physicians knew how to deprive a cancer cell of its ability to travel, which they call metastasizing, they would have a powerful new weapon against a great scourge. (We'll return to cancer in Chapter 12.)

There is also an implicit reason why cell biologists pursue the mysteries of cell motility. Autonomous motion, as we saw in previous chapters, is one of the classical hallmarks of life. It is one of the properties that sets the cell apart from any other structure in the universe. Only the living cell can pick itself up and travel to a new location. At a deep level, this phenomenon reveals that the cell is also the fundamental unit of behavior. The cell exhibits the fundamental unit of volition, of will—the power to respond to the surrounding environment through a physical act of behavior. The macrophage that approaches the bacterium and eats it, the embryonic cell that follows some chemical signpost to another part of the embryo, the metastasizing cancer cell—all are mere collections of molecules obeying the laws of physics and chemistry. This is what gives them the power to move. This is a large part of what gives them life.

THE LIVING-ROOM CELL

CHAPTER **4**

Imagine that the room in which you are sitting is one gigantic cell. You are inside the cell, sitting comfortably, and simply by looking around the room you can study the internal workings of this improbably huge cell. If you are in an average-sized living room, the nucleus (which holds the genetic blueprints in the form of the chemical called DNA) would be nearly the size of a Volkswagen Beetle—parked, let's say, over against the wall to your left.

This metaphor of the living-room cell will arise several times in this and the next chapter because it is easier to think about the extraordinary web of objects and processes within any cell if they are scaled up to dimensions that are familiar in the everyday world. After all, typical human cells are really so small that 250 could nestle side by side on top of the period at the end of this sentence.

Take a close look at the tiny ridges of skin that make up your fingerprints. Each ridge is about twenty cells wide. A patch of skin measuring just one square inch has more than one million cells in the top layer alone. As it happens, not all cells are so small. The record holder is the ostrich egg, which is somewhat bigger than a grapefruit and, like all eggs, consists of a single cell. Even within the human body there are some respectably large cells. Muscle cells are long, thin filaments that can stretch an inch and a half. Still longer are the nerve cells that extend out from the spinal cord to the tips of the fingers or toes.

Still, the size of the vast majority of cells—certainly most of those in the human body—is so tiny that it is very hard to think about what goes on inside them while imagining them at their true, sub-Lilliputian scale. Cell biologists can get around this by using powerful microscopes to see cells, microscopes with such magnifying power that a photograph of the inside of one cell taken through such a microscope can show details as clearly as an ordinary snapshot of your living room. The imaginary living-room cell—roughly 300,000 times bigger than a typical real cell—allows one to study cells with a kind of mental microscope—a visual metaphor that makes it easier to understand how life works. This is so because most of what happens in cells involves physical objects interacting in ways that take many words to describe but which really are no harder to understand than how Lego blocks fit together or why each size of bolt requires a special wrench to turn it. If you can visualize Lego blocks or wrenches, it's a lot easier to understand how they work than if you read a description of them in words.

Like all cells, the living-room cell is jammed full of organelles from wall to wall, floor to ceiling. In the living-room cell the force of gravity is negligible, as it is in real cells, and many objects float as if they were weightless in space; others are stuck to various surfaces and move with them. Along with the huge VW-sized nucleus there are half a dozen floppy stacks of beanbag chairs undulating softly in the air. The "air," of course, is the water inside a cell, which is thick with all the molecules and other structures in it, so thick that it is viscous, or jellylike.

From time to time, a little bit of a beanbag pinches off to form a bubble about the size of a golf ball. It drifts slowly, like a soap bubble, bumps into a long rope that reaches off to the wall, and suddenly begins gliding along the rope, headed for the far wall. That was a vesicle being transported by a motor molecule, probably the kinesin we discovered in Chapter 2. There is a dense web of ropes and strings running every which way throughout the room. Some are straight. Others look more like a filigree, branching like the limbs of a tree. Some strings seem to wrap around the nucleus and then stretch off to attachment points on the walls. Other ropes line the walls, pinned to it at intervals.

Scores of long sausages slither about. Some of them also seem to glide along ropes, also carried by motor molecules. There are vast areas of great floppy sheets, like a deflated hot air balloon that has been folded loosely and studded with thousands of marbles. The layers and layers of balloon wrap around the VW nucleus, almost concealing it. The strings around the nucleus pick their way through holes in the sheets or between them. And there are hundreds of grapefruits and some basketballs hovering all through the room.

In a real cell, all this stuff—biologists refer to it collectively, except for the nucleus, as *cytoplasm*—would be so thickly strewn that an armchair biologist surveying it from inside wouldn't be able to see very far. But, fortunately, the mental microscope allows us to see through structures that are not of immediate interest. We'll revisit all of these organelles in due course.

Look over at the nearest wall, at the cell membrane, also called the *plasma membrane*. This membrane encloses the entire cell, like skin that encloses our bodies, and forms the outer boundary of every cell. The cell membrane, no thicker than shoe leather at this scale, is one of the most important and most active organelles. At thousands of places across its surface, it holds gatekeeper structures—receptors, pores, channels, and others—each specialized to take in or put out specific molecules (Figure 4.1). In a cell that lines the intestine, for example, the membrane's gatekeepers decide which products of digestion are absorbed and relayed to the bloodstream. The membranes of certain white blood cells (which are parts of the immune system) take in AIDS viruses because those microbes wear a disguise that tricks the membrane into thinking the virus is friendly and should be welcomed into the cell. The plasma membranes of all cells control the movement of nutrients, hormones, and other molecules into the cell. The same membranes also regulate the export of wastes and newly made substances from the cell into the surrounding region. In certain cells of a woman's breast, for example, it is this membrane that regulates the secretion of milk into the ductwork that carries it to the nipple, and in the pancreas, these membranes control the release of insulin into the bloodstream.

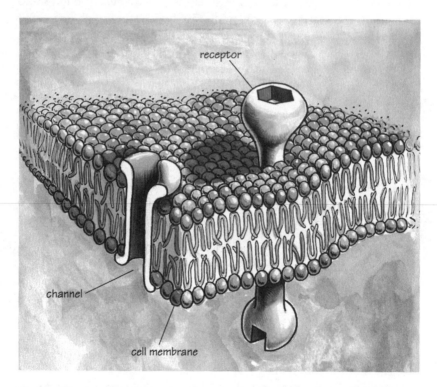

**Figure 4.1** Protruding through the outer membrane of every cell are thousands of structures that act as gatekeepers, controlling the flow of specific molecules into or out of the cell. Some are receptors that, upon binding to a specific incoming shape, activate some process within the cell or physically pull the molecule into the cell. Other gatekeepers are "channels," "ports," and "pumps" that govern the passage of small molecules.

An understanding of how membranes work has led to several dramatic advances in medicine, including development of "beta blockers" to treat heart conditions; the drug lovastatin, which reduces blood cholesterol; and an experimental AIDS drug that links a cell-killing chemical to a molecule that binds only to the membrane of cells that have made the fatal mistake of taking in an AIDS virus. We'll examine these after having a closer look at the membrane's basic structure.

The main part of the cell membrane is made of countless trillions of small, double-ended molecules called phospholipids. The detailed chemistry of phospholipids endows them with extraordinary properties, without which life would be impossible. For example, the head end is a phosphate molecule that has the chemical property of attracting water and is, therefore, called hydrophilic ("water-loving," in Greek). The tail end is a fatty material, or lipid. Lipids are water-repelling, or hydrophobic. (As everyone knows, oil, which is made of lipids, and water don't mix.)

Put a batch of phospholipid molecules in water and an amazing thing happens. The molecules spontaneously organize themselves into a two-layered sheet, with the water-attracting heads on the outside, where they touch water, and the water repelling tails on the inside, away from contact with the water. The tails of each layer touch those of the other (Figure 4.2).

The sheets will even curl themselves into hollow, two-layered balls filled with water, so that no sheet is left with an edge where the hydrophobic tails can get wet. Although the membrane is microscopically thin, it turns out that the attraction among the phosphate heads is quite strong, producing a membrane that is not easily ruptured. The plasma membrane is made even stronger by the addition of *cholesterol*—the same stuff that nutritionists say we should eat less of. Although an excess in the bloodstream is not good, cholesterol is essential to all living cells. Because cholesterol is a lipid with one end hydrophobic and the other hydrophilic, molecules of it slip easily in among the phospholipids. The more cholesterol in a cell's membrane, the stronger it becomes. Scientists who inject cells, for example, routinely see the ultrasharp tips of their glass needles push against a cell membrane, causing it to dimple deep into the cell, like a finger pushing into an inflated rubber balloon, before the needle breaks through. When that happens, the cell does not—unlike the balloon—pop. The membrane stays intact and seals itself instantly when the needle is withdrawn, restoring its integrity. This remarkable property of the cell membrane allows a wide range of experiments in which cell biologists inject individual cells with various substances either to alter their structure and behavior or to stain specific organelles so that they can be tracked under the microscope.

Every cell of every living thing on Earth, and the organelles in those cells, is enclosed by this kind of two-layered phospholipid membrane.

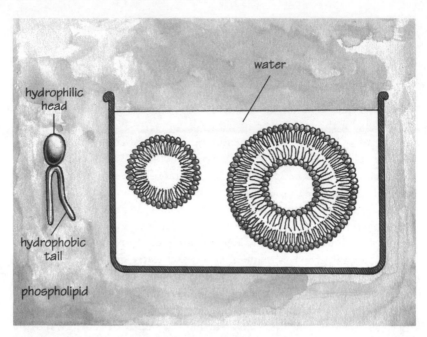

**Figure 4.2**  All membranes in cells consist of two layers made up of countless identical molecules called phospholipids. This molecule's chemical proclivities cause it to behave predictably in water, forming sheets or closed bubbles. This happens because the dangling tails (lipids) are water repellent and attractive to one another, while the heads (phosphates) are attractive to water.

(Bacteria, a simple form of life whose cells are quite different from those of all multicellular organisms, have no discrete organelles and, hence, no membrane-bound organelles inside their bodies, but they are enclosed by the same kind of outer membrane.)

Most of the substances that play prominent roles in life cannot get through the membrane without the aid of a specific structure, a gate-keeper stationed in the membrane. Some of these structures are the receptor proteins that we have already met. Others are simpler affairs, called *channels*, which allow small molecules, including water, to diffuse into or out of a cell. Channels are holes in the membrane, each lined with protein molecules to keep the membrane from squeezing the hole shut. In the wall of the living-room cell, they would be no bigger than a nail hole through the quarter-inch thickness of the membrane.

Some membrane structures are more complicated affairs and work in different ways, each specialized to allow passage to a specific class of molecules. There are ports that take in nutrients; there are pumps that haul in or throw out electrically charged atoms such as calcium, sodium, and potassium; and, of course, there are the receptors, structures that seem to grow in importance every year in explaining how life works. Some receptors literally grab certain passing molecules that the cell "wants," yank them through the membrane, and release them inside. Other receptors keep the arriving messenger molecule outside the cell but themselves undergo a change in the part of them that sticks inside the cell. That change may trigger some new event, such as releasing or activating a second messenger molecule, which goes on to do some job. Thus the arriving molecule can act as a signal whose message is transduced through the membrane and relayed not as the original signal itself but in some other form—much as a microphone is a transducer, converting sound waves into electrical signals.

The study of second-messenger systems in cells has led to an understanding of why cholera, one of the historic plagues of the Third World, is such a deadly disease. Cholera is caused by certain bacteria that are spread when sewage water containing raw human waste mixes with drinking water. If a certain molecule produced by the cholera bacterium gets into cells lining the intestine, it triggers a series of chemical reactions that ultimately cause one particular type of receptor to stay locked in the "on" position. In this case, the mechanism that is turned on causes the cell to secrete water, pouring it into the intestinal tract. When all the cells lining the gut do this, the result is severe diarrhea. Cholera kills chiefly by removing so much fluid and dissolved substances from the blood that victims are killed by severe imbalance of electrolytes (charged atoms such as sodium and potassium) and dehydration.

In the living-room cell, each of these gatekeeper structures is no bigger than a nail head in the cell's membrane. Each type of cell in the body has its own complement of receptors, channels, pumps, and other ports of entry and exit, the number and type varying with the cell's specialized needs. Tens of thousands of receptors, sometimes hundreds of thousands, are embedded in the outer membranes of most cells. All these structures are held in the membrane because they, like the membrane, have hydrophobic and hydrophilic regions. A typical receptor protein, for example, can be thought of as an irregularly shaped structure with hydrophilic regions and hydrophobic regions. Put this molecule next to a biological membrane and the ordinary chemical forces it encounters will pull it, like a nail next to a magnet, into position. The hydrophobic parts will wind up deep inside the membrane next to the membrane's hydrophobic parts, thus anchoring the molecule in place, and the water-attracting parts will stick out on either side. Some receptors have alternating hydrophilic and hydrophobic regions that make the molecule snake back and forth through the membrane several times.

Unlike your living-room wall, large regions of the cell membrane are fluid, at least in a sense. They are not rigid. Drive a nail into your wall and it stays put. But thrust a receptor into the cell membrane and it will drift around, something like a block of wood floating on the surface of a pond. Biologists speak of the membrane as a "fluid mosaic." In other words, the phospholipids are mosaic tiles, always staying side by side in the same plane, but they can slip around one another. And anything embedded in them can slip too, something like the way a determined pedestrian can slip through a shoulder-to-shoulder crowd without leaving the plane of the ground.

Heart muscle cells have receptors for the hormone epinephrine, a stimulant that is also called adrenalin. This is the "fight-or-flight" hormone that a body produces in extra amounts when under stress. One effect is to speed the heartbeat. But a fast heartbeat can be dangerous in people with heart disease. Researchers have developed a molecule that is shaped something like the stimulant molecule—enough like it that it can bind to the adrenalin receptor. But, unlike adrenalin, it fails to trigger the second-messenger system within heart cells. Thus it does not simulate the heart. Instead, by occupying the receptor, it keeps adrenalin away. The receptor is called a beta adrenergic receptor, and the drug, a beta blocker. Beta blockers are now used by millions of people with heart trouble and even by perfectly healthy musicians, public speakers, and other performers who suffer the adverse effects of the excessive adrenalin production that sometimes accompanies stage fright. By disabling some of their adrenalin receptors, these performers stay calm and controlled.

Lovastatin, a drug that reduces cholesterol in the bloodstream, also works through receptors on the cell membrane. The cholesterol that cells need is acquired both by extraction from the bloodstream—via cholesterol receptors—and by manufacture within the cell. Lovastatin is a molecule that blocks a key step in the cell's process of making cholesterol. When the cell senses that its own ability to make cholesterol isn't working, it compensates by making extra cholesterol receptors and installing them in its outer membrane to pull more cholesterol out of the bloodstream. The result is less cholesterol in the blood to deposit on artery walls.

The discovery that cells can behave this way earned two University of Texas researchers—Michael Brown and Joseph Goldstein—a Nobel prize in 1985. Their work began in the early 1970s when they independently became interested in the same rare disease, called familial hypercholesterolemia, and eventually they became close collaborators. The disease is a hereditary affliction that comes in two forms. In the mild form, its victims die of heart attacks in their forties or fifties. In the severe form, they succumb in their teens or even younger. Heart attacks have felled even six-year-olds with the disease.

The problem is that sufferers have massively high amounts of cholesterol in their bloodstreams. The excess cholesterol forms thick deposits on artery walls, progressively narrowing them until a blood clot lodges in the narrowed opening and blocks further blood flow. This can be life-threatening when the affected vessels are the coronary arteries, the ones that supply blood, with its vital cargo of oxygen, to the heart muscle. If the coronary arteries are blocked, the result is a heart attack (or coronary infarction), and the affected region of muscle suffocates and dies. This can happen to almost any genetically normal person, especially late in life, but to people who have this disease, it happens at much younger ages. The mild form was known to occur in about one out of every 500 people worldwide. The severe form is much rarer, afflicting about one in a million.

Brown and Goldstein discovered that cells take in cholesterol by using receptors not for cholesterol itself but for another molecule that carries cholesterol in the blood. The carrier is called low-density lipoprotein, or LDL. Though sometimes described as the "bad" cholesterol, LDL is not cholesterol; it is a carrier of cholesterol. (It's called "bad" because in the course of delivering cholesterol to cells that need it, it also deposits some on artery walls, where it can build up and clog the artery. Its "good" counterpart is high density lipoprotein, or HDL, which removes cholesterol from artery walls.)

The two researchers found not only that cells have LDL receptors. They also learned that the receptors work in an extraordinary way—a way that reveals yet another case in which mindless molecules contain all the information and ability needed to carry out acts that seem to have clear intent and profound purpose. Like all receptors, those for LDL are proteins whose extracellular part—the part that sticks out of the cell—is shaped to fit the complementary shape of an LDL molecule. The receptors also have a part that sticks inside the cell, and when the outside part binds to an LDL (carrying its cholesterol molecule), the inside part changes its own shape, turning itself into a form that can bind to yet another protein that is drifting inside the cell.

This other protein, called *clathrin,* usually exists in the form of three rods attached to a hub, like the star in the Mercedes-Benz emblem. Under the right conditions, isolated clathrin stars will link spontaneously at the free ends of the rods, forming the correct angles to make the bars of a cage, arranged as a mosaic of interlinked pentagons and hexagons identical to the geodesic dome attributed to the architect Buckminster Fuller. As more clathrin stars join the dome, they eventually close it into a geodesic sphere. The genius of the geodesic dome, and the sphere, is its ability to make a curved, three-dimensional structure out of simple straight rods—a design that also happens to be mechanically very strong.

Clathrins do not do this, however, unless they are bound to the tails of receptor molecules sticking inside the cell. And those tails must first have taken on the altered shape caused when their external parts grabbed onto LDL molecules. In the living-room cell you can just barely make out these tails sticking in from the walls like bits of string. As the receptors grab

LDLs outside the cell, each tail inside the cell links up with a tri-star clathrin. Then, the LDL receptors somehow begin clustering together. This brings the clathrins close enough to link to one another and begin making their little geodesic dome. The clathrin–clathrin bonds are so strong—trying, so to speak, to make a geodesic sphere—that they deform the surface of the cell membrane, sucking the pits deeper into the cell. From the living-room armchair, it looks as if somebody outside the room is trying to push a golf ball through the cell membrane.

More LDL receptors move across the cell surface to join the complex, presumably drifting aimlessly through the fluid mosaic of the cell membrane until their clathrins bump into other clathrins and lock. Gradually the pit deepens. The golf balls push into the cytoplasm of the living room, eventually forming a pouch. From the outside the pit can be seen to deepen. Then, as the membrane closes over the hole, the pit—now a bubble, or vesicle—breaks free and drifts into the room (Figure 4.3).

On the outside of the ball is the geodesic cage made of clathrin. Just inside that is a hollow ball of membrane wrenched from its position in the

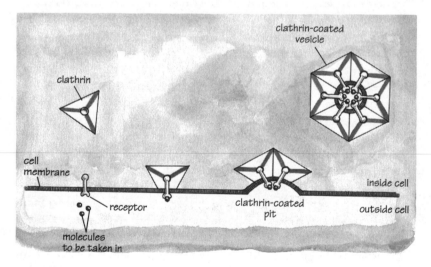

**Figure 4.3.a**

**Figure 4.3**  (a) Many molecules are taken into cells through the action of so-called clathrin-coated pits. When the molecule binds to the receptor, it changes the receptor's intracellular tail in a way that allows clathrin to bind. Bound clathrins spontaneously link to one another in a way that forces the membrane to dimple inward. As more and more clathrin molecules link up, the pit deepens and breaks free inside the cell as a vesicle. The incoming molecules (circles) are now inside the cell. (b) Real clathrin-coated pits as seen under a microscope. (Reproduced from *Journal of Cell Biology* 84 [1980]: 560, by copyright permission of the Rockefeller University Press. Photo courtesy of Professor John Heuser)

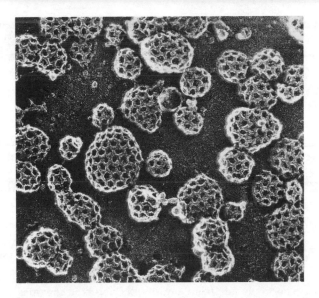

**Figure 4.3b**

cell's skin, but still studded with LDL receptors. At the center of the ball are the LDLs themselves, each carrying many cholesterol molecules. The molecules that once were outside the cell are now inside but contained in a vesicle in which they will be transported for further processing.

Brown and Goldstein's Nobel came not just for discovering all this but also for explaining how this mechanism is involved in familial hypercholesterolemia. They showed that if people have defective LDL receptors, their cells take in too little cholesterol, leaving so much of it in the bloodstream that their arteries clog up early in life. Their LDL receptors are defective because the people have inherited flawed genes for the receptor, a subject we'll take up in the next chapter, which concentrates on how genes govern life.

Many cell biologists suspect that the mechanism of importing molecules with clathrin-coated pits is the dominant method used by cells to take in outside molecules. It has been proven, for example, that cells take in insulin this way, as well as several growth factors, which are molecules that stimulate the cells to multiply.

Every cell in the human body must, like any organism, consume food to live. The cell membrane, among its many roles, is also the cell's mouth. For example, macrophages, the amoebalike creatures we met in Chapter 3, whose main role in the body is eating, gobble up bacteria and other invaders as well as worn-out cells. When a macrophage engulfs a morsel of food, it gradually embraces it with its own cell membrane. The enclosed volume then is inside a bubble of membrane—a vesicle, or *endosome,* as it is often called—that breaks away from the outer surface of the cell and drifts inward. The plasma membrane simultaneously seals itself, as if your fingers closed around a coin in the palm of your hand and the skin of your fingers and palm fused. The coin would now be inside your body.

When a macrophage has completely enveloped a morsel of food, the vesicle containing it moves inward to meet one of the cell's many stomachs, organelles called *lysosomes,* from the Greek *lysis* meaning to "loosen or break apart," and the Greek *soma* meaning "body." In the living-room cell there are about a hundred or so grapefruit-sized to basketball-sized lysosomes hanging about, each enclosed by a membrane similar to the one that surrounds the entire cell. Inside each lysosome are about fifty different kinds of digestive enzymes, a corrosive brew so powerful that if a lysosome ruptured and spilled its contents, the cell would fatally digest itself away. Some of these enzymes break down proteins, others attack fats, and still others disassemble the nucleic acids that make up DNA and RNA.

Cells are normally safe from a mishap because the enzymes work only in an acidic environment, with a pH of 4.8 or less (that's about the same acidity as tomato juice). In the cytoplasm in general, the pH is neutral, 7.0, to slightly alkaline, 7.3, or about the same as tap water. This form of protection, however, can break down. Suffocation or drowning can be so quickly fatal because the lack of oxygen causes cells to become more acidic. If the pH drops low enough, the lysosomal membranes break down, releasing the enzymes to destroy the cell from inside. Brain cells are the first to undergo this reaction, often destroying themselves in four to five minutes after breathing stops. The rate of this destruction, like all chemical reactions, depends on temperature—the colder it is, the slower the reaction. Chilling also slows metabolism, thus slowing oxygen consumption in the cell. This is why people who fall into icy water may survive for half an hour or more without breathing.

Looking out at the wall of your living-room cell—which for now is a macrophage—you see a salami-sized blob move deeper into the room, carrying a hapless bacterium until it bumps into a lysosome. The membranes of the two structures quickly fuse, merging like two drops of water becoming one, in a controlled process. The membranes of each organelle—fuser and fusee—must carry opposite members of a complementary set of *docking proteins* or else all the organelles would quickly merge into one giant blob. Only after they lock do the membranes fuse, pitching the bacterium into the lysosome's digestive juices, where it is broken down to mere molecules.

The process of digestion is much like disassembling a chain by breaking links. This is so because many of life's fundamental molecules are polymers—long molecules formed by repeatedly linking, smaller subunits, or monomers, into long chains. For example, a polymer of protein is made up of linked monomers called amino acids; DNA and RNA, the heredity-carrying molecules, of linked nucleotides; and carbohydrates such as starches, of linked sugars. Lysosomes contain digestive enzymes specialized to break apart all of these polymers. They also have enzymes to break down fats, which are not polymers, into subunits called fatty acids.

Before long, the lysosome that received the bacterium contains only soup, a rich broth of subunits, raw materials that the cell will use to make new large molecules tailored to its own specifications.

Most other human cells, unlike macrophages, never dine on such exotic fare as bacteria. More often, their lysosomes are recycling centers that break down worn-out organelles from within the same cell or that degrade incompletely digested nutrient molecules delivered from the gut by the bloodstream.

The clathrin-coated vesicles that contain receptors bearing LDLs linked to cholesterols are also destined for processing in the lysosome. First the clathrin cage is removed from the vesicle (and the clathrin tri-stars are recycled to make new vesicles). Then the receptors in the vesicles let go of their LDL-cholesterols. The receptors gather at one side of the vesicle, which soon buds off from the main vesicle, leaving the loosed LDL-cholesterols behind. The vesicle with the released receptors heads back to the cell surface, fuses with it, and reinstalls the receptors for reuse. The vesicle left behind with the LDL-cholesterol fuses with a lysosome, dumping its contents into the digestive brew. The LDL is broken into amino acids to feed the cell, and the now-freed cholesterol is available for use in the cell's membranes.

Lysosomes don't always work right. When they fail, the result can be catastrophic disease. In Tay-Sachs disease, for example, victims have inherited a faulty gene, which leads to a defective enzyme or no enzyme at all. As a result certain undigested molecules accumulate in the lysosome, gradually leading to a kind of lysosomal constipation that makes the organelle swell so big it essentially chokes the nerve cell to death. Before long, so many nerve cells are destroyed that the brain is unable to function normally. In Tay-Sachs disease the disruption of brain function eventually becomes so severe that its victims often die in early childhood. About two dozen other so-called lysosomal storage diseases are known.

Most other cells of the body, unlike macrophages, cannot chase down food. They must rely on a delivery service—the bloodstream and its rich broth of nutrients from the digestive tract. Because the human stomach contains its own enzymes for breaking down proteins, carbohydrates, and fats, most cells simply take in already-digested nutrients in the form of amino acids, glucose, and fatty acids. The bloodstream delivers the nutrients through a vast network of blood vessels that branch repeatedly into smaller and smaller vessels to reach as close as possible to every cell. The smallest vessels, the capillaries, are so thickly distributed through most tissues that no cell is more than three or four cells away from nutrient-rich blood. Some of the nutrients reach distant cells because they leak out of capillaries and flow in the tiny spaces between nearby cells.

But there is also a more active delivery service. Imagine for a moment that your living-room cell is one of those that make up the lining of a capillary. To the left of where you sit, just outside the membrane, is the open space of the blood vessel, where blood flows past. To the right are hungry cells not in contact with the blood supply. Your living-room cell must act as an intermediary. The wall on the left, on the blood side, suddenly appears to push into the room, forming a bulge the size of a grapefruit. The bulge swells, protruding farther into the room until it breaks through, allowing the membrane-enclosed bubble to float into the room. The bubble, or *endosome,* containing nutrients from the blood-stream, moves across the room to the opposite wall, passing by your own lysosomes, and fuses with the membrane on that side, discharging its contents into the space just outside your cell, where the next cell can get it. The nutrients diffuse through the narrow spaces between cells to reach other cells that are off the beaten blood path.

Most human cells take in their food not by making endosomes but through yet another kind of gatekeeper molecule, or port, embedded in the plasma membrane. There is, for example, the glucose port. This is a protein whose outer end, the part sticking outside the cell, has binding sites for both a sodium atom and a glucose molecule. The sodium is essential because until it binds, the port does nothing. When it does, it makes the protein change shape while still embedded in the membrane, swiveling so that it drags the glucose through the membrane and into the cell and releases it. It also releases the sodium atom inside the cell. The cell has just "eaten" one molecule of glucose sugar and one atom of sodium.

Cells also have amino acid ports—different kinds for different amino acids—that work the same way. Each time one of these ports grabs an amino acid and yanks it into the cell, the requisite sodium atom comes in, too.

The glucose and the amino acids go on to become sources of energy and building blocks for the cell's growth and development. But not the sodium atoms. They can be a problem for the cell, and to prevent sodium atoms from accumulating to dangerous concentrations, cells have *sodium pumps* to throw them out as fast as they come in. As it happens, when the sodium pump tosses out a sodium, it tends to bring a potassium atom into the cell at the same time. Actually, for every three sodiums pitched out, two potassiums are pulled in. This takes care of an opposite problem, the fact that cells are constantly losing potassium and must replace the losses. These devices are called pumps because they must work against a resisting force, moving atoms "uphill" in a chemical sense. Normally the sodium concentration outside a cell is higher than inside. Thus osmosis would tend to drive sodium inside until the two concentrations were equal. The potassium situation is exactly the opposite. The sodium-potassium pump consumes some of the cell's energy in order to push these atoms in opposite directions, both chemically uphill.

This sodium-potassium pump helps, along with the similar calcium pump, with another essential job—that of maintaining the cell in the proper electrical state. Life, it turns out, is more than a chemical phenomenon. Life is also electrical. Every cell carries its own electrical charge and uses it to perform a variety of tasks. Rapid changes in charge are essential to the ability of nerve cells to transmit signals, to the contraction of muscle cells, to the stimulation of secretory cells in glands. It is even essential to the union of sperm and egg at conception. The neurologist's EEG (electroencephalogram) is produced by a kind of radio receiver that records the electromagnetic waves given off by the brain as billions of brain cells discharge their electrical impulses to one another. The EKG, or ECG, (electrocardiogram) is the cardiologist's equivalent for recording heart activity. Both devices can help diagnose illnesses in which the electrical properties of the relevant organs are abnormal. If the cells are malfunctioning, their electrical properties will be out of whack.

Electrical conduction, moreover, keeps the heart beating in a coordinated rhythm. Problems with the electrical balance of heart cells can lead to fatal arrhythmias, conditions in which the heart cells have lost the ability to regulate their electrical charge and can beat only in a fluttering, uncoordinated way. Such arrhythmias have killed many people, including some young athletes seemingly in the prime of health. Presumably if their sodium-potassium pumps were working better, they would not have been at risk.

Glucose, the sugar that enters the cell via the glucose port, is the cell's main source of energy. Its power is stored, like that in any other molecule, in the form of the chemical bonds that hold its various atoms together. As long as the atoms are bound, they contain the energy between them. If the atoms are split apart, the energy is released.

Glucose is rich in high-energy chemical bonds. They were installed there by plant cells performing photosynthesis. In that process, molecules of carbon dioxide (carbon and oxygen) and water (hydrogen and oxygen) are taken apart and reassembled into glucose (carbon, hydrogen, and oxygen). The breaking down and reassembling are powered by solar energy, some of which is captured in the glucose molecule's bonds. In the process, fortunately for animals, there is some leftover oxygen, a waste product that plants excrete into the air. Human cells, like those of all animals and plants, reverse photosynthesis by taking glucose apart and releasing the energy to power their own growth and development. This process also gives off two by-products, water and carbon dioxide, the original ingredients consumed by the plants. To carry out these chemical reactions, animal cells also need some additional oxygen atoms. These diffuse in from the bloodstream, which got them, of course, in the lungs.

Glucose energy, however, is not in a form that the cell can use directly. Energy must first be extracted from glucose and processed into another form. The first part of the job happens in the open spaces in the cell—in the narrow, watery regions between the organelles. In the living-room cell you couldn't see glucose molecules very well. They were mere dust specks. But you could see some of the enzymes that act on them—little blobs ranging in size from BBs to grains of sand. They are catalysts, making the reactions happen but not themselves being changed in the process. After nine successive chemical reactions, each performed by its own enzyme, one glucose molecule (a chain of six carbons fringed by twelve hydrogens and six oxygens) is broken down and its atoms rearranged into two molecules of a substance called pyruvate (a three-carbon chain with three hydrogens and three oxygens).

The pyruvate, still carrying the stored solar energy, is taken into yet another organelle floating in the living-room cell—those long, undulating sausages that cell biologists call *mitochondria,* from the Greek terms *mito* meaning "thread" and *chondrion* meaning "granule" (Figure 4.4). These terms, incidentally, were assigned long before anybody knew the functions of specific organelles and are, at best, crudely descriptive of what could be seen with microscopes of limited means. Today it is known that mitochondria are the cell's powerhouses, the living equivalent of an electrical generating station. Just as the power company takes various forms of energy (coal, oil, gas, nuclear, etc.) and converts it into one convenient form that everyone can use (electricity), mitochondria take the energy from digested food—the pyruvate—and convert it into a form that all parts of the cell can use.

Though there are only a few hundred mitochondria in the living room, some cells with higher energy demands, such as muscle cells, have thousands. In heart muscle cells, for example, about half the volume is occupied by mitochondria, which are needed to supply the enormous energy demands of cells that must perform large amounts of work.

Inside the membrane enclosing each mitochondrion is a second membrane that is deeply and repeatedly pleated. You can see this in a living-room mitochondrion because the outer membrane is translucent and inside there is what looks like a casing for a much larger sausage, only it's collapsed into a pile of wrinkles so that it can fit inside the smaller membrane. The point of this curious arrangement is to pack the largest possible surface area inside the outer membrane. The reason is that the inner surface is the site of the key chemical reactions that process every cell's energy supply.

Though the chemistry of what goes on inside a mitochondrion is

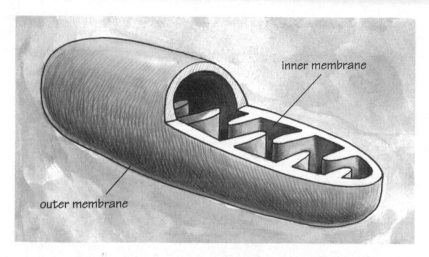

Figure 4.4a

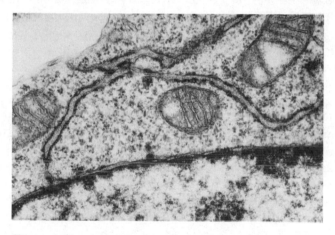

Figure 4.4b

**Figure 4.4** Mitochondria are the organelles in which various forms of chemical energy taken in by the cell are converted to one form that can be used throughout the cell. That form is ATP, adenosine triphosphate. (a) The diagram shows the typical form, which includes a deeply folded internal membrane that produces a huge membranous surface area inside the mitochondrion. (b) The micrograph, taken with an electron microscope, shows real mitochondria. In reality, they are very long but because the picture is of a thin slice through a cell, some mitochondria have been sliced more lengthwise, others, more crosswise. The large curve at the bottom of the picture is the outer membrane of the nucleus, containing the DNA. Double lines undulating through the image are endoplasmic reticulum, or ER, which will play a key role in Chapter 5.

complex and a full description is beyond this book, it is well understood by biochemists. Defects in some mitochondrial processes can lead to several diseases of muscles called mitochondrial myopathies. Some researchers suspect that heart failure, a condition in which the heart cannot pump enough blood to meet the body's needs, may be the result of defective mitochondria failing to supply enough energy to the heart muscle.

Inside the mitochondria, pyruvate is processed to release its energy, which is in the form of high-energy electrons, the same subatomic particles, as it happens, that make for electricity. In this case, however, the electrons do not flow through a wire. Instead, they are immediately recaptured to make new chemical bonds in new molecules. After another complex series of chemical reactions—including the famous Krebs cycle, a loop of reactions discovered by Sir Hans Krebs, a British-German biochemist, in the 1930s—the electrons liberated from the pyruvate are eventually recaptured to make the molecule that is the cell's nearly universal energy source—ATP, or *adenosine triphosphate*. ATP is the immediate source of virtually all living energy in all cells of all organisms on Earth. ATP powers the molecular motors that carry vesicles within cells and it powers the muscular contractions that allow the body to move. Nerve cells use the energy in ATP to make electricity for nerve conduction. Some of ATP's energy is also released as the heat that keeps warm-blooded animals alive.

All this happens because each molecule of ATP packs a tremendous wallop of concentrated solar energy, captured originally by green plants performing photosynthesis and now stored in the form of high-energy electrons. Some chemical reactions in cells occur spontaneously because of the inherent chemical properties of the molecules involved. Many other reactions, however, will not happen unless there is some outside source of energy. These reactions are like watches or transistor radios that need batteries to operate. Mitochondria, then, make ATP batteries and ship them out to all other parts of the cell. When a battery runs down, as each molecule of ATP does once it is used, it is sent back to a mitochondrion for recharging.

ATP is a simple battery. It consists of one molecule, called adenosine, to which are attached three phosphate groups end to end, hence "adenosine triphosphate." The energy is in the bond that links each phosphate to the remainder of the molecule. ATP batteries are dispatched through the mitochondrial membrane with the aid of gatekeeper molecules. (Unfortunately, they're too small to see in the living-room cell.) The batteries diffuse throughout the cell. If they happen to bump into a pair of molecules that would react if they had the energy, ATP instantly sheds one phosphate, releasing the energy that held it, and becomes *adenosine diphosphate,* or ADP. The powered reaction takes place, where, for example, two molecules may combine or an enzyme may chop another molecule in two. The ADP, now in need of recharging, drifts until it finds a mitochondrion and is taken in to be outfitted with another phosphate.

Mitochondria are one of the more unusual organelles in the cell because there is good reason to believe that they once had a life of their own. The hundreds of mitochondria slithering about inside each of your cells are, in fact, very much like bacteria that have permanently colonized more complex cells. There is considerable evidence: Mitochondria are the same size as bacteria, and each mitochodrion has its own DNA—a little chromosome that carries several genes. As in bacteria, the length of DNA is closed into a loop. (The chromosomes in the cell's nucleus, by contrast, are strands with two unattached ends.) Mitochondria have machinery for reading and carrying out the instructions of their own genes, and they reproduce by splitting in two—just like bacteria. In other words, mitochondria are not manufactured in quite the way other organelles are. They multiply on their own, and every time a cell divides, each daughter cell acquires a small colony of them.

Such observations long ago led early cell biologists to suggest that mitochondria are the evolutionary descendants of bacteria that took up permanent residence in some primordial cell. Either they were symbiotic inhabitants of the cell or the cell tried to eat the bacteria, which somehow resisted digestion. The result was an extraordinarily intimate form of symbiosis, one that conferred such profound advantages on the host cell that that original cell evolved into all the higher forms of life on Earth, from yeasts through human beings. Those once-independent bacteria haven't done too badly, either.

This so-called endosymbiosis theory, once regarded as somewhat fanciful, gained general acceptance in recent years after Lynn Margulis of the University of Massachusetts and others developed additional evidence to support it and after Margulis began promoting it in her writings. Most cell biologists now think the theory is true, but as yet, it does not fully explain some important points. For example, the DNA in mitochondria does not contain anywhere near the number of genes needed to make new mitochondria. The additional genes are stored among the huge complement of cellular genes in the nucleus—the VW Beetle. If the mitochondrion began life as a bacterium, somewhere along the line it must have transferred most of its genes into the nucleus, surrendering much of its essence as a whole, independent organism to its host's master control center.

Still, there is one other point of similarity between mitochondria and free-living organisms. Eventually, it seems, they get old, wear out, or break down. Biologists don't really understand what happens, but it looks as if mitochondria die and are, in a sense, eaten by their host cell, a process biologists call autophagy. They infer this from the fact that inside some lysosomes—those cellular stomachs full of digestive juices—recognizable

bits and pieces of mitochondria have been found. In time the mitochondrion is completely digested away, reduced to its raw materials,
which will become building blocks for yet another set of processes within
the cell—the manufacture of new molecules and new organelles to keep
life going, the subject of the next chapter.

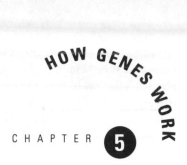

**A** little boy races across a soccer field, intercepts the ball, and kicks with all his might. Minutes later he is writhing on the ground in tears, his legs burning with pain. He does not know it, but he could die from what is happening inside the cells of his legs.

His exertion has triggered chemical and physical reactions that with deadly swiftness are turning his normally life-giving red blood cells into killers. The usually disc-shaped, round-edged, and pliable oxygen-carriers are being wrenched and twisted into the shapes of crescents, or sickles. Like nails trying to flow through a garden hose, they jam. In the narrow capillaries, only wide enough to admit red blood cells single file, the sickled cells stop flowing altogether, blocking the delivery of oxygen to tissues. The cells of the child's leg muscle, because they have just run up an oxygen deficit by working so hard, are the first to succumb. Because the blood cannot flow fast enough, the muscle cells are not getting enough oxygen. They are suffocating and they begin to die.

Along with 50,000 other Americans, the boy has inherited sickle-cell anemia, one of the most common genetic diseases in the United States. This severe and often fatal affliction is caused by the tiniest of all possible flaws in the genes—a single erroneous "letter" in the 861 letters of genetic code that dictate the structure of hemoglobin, the protein inside red blood cells that carries oxygen. Those 861 letters constitute a gene for hemoglobin—one genetic "sentence" that a cell's manufacturing machinery can read and follow to make hemoglobin. Like the hemoglobin gene, all genes direct cells to make but one type of molecule, a protein. Proteins, as we will soon see in greater detail, are long chains, or polymers, of subunits called amino acids. Proteins function as structural elements (the bricks and mortar of cells) and as enzymes (the mechanics that take apart other molecules or join them together).

The genetic misspelling error that causes sickle-cell anemia has no significant effect on a person's health under most conditions. But if the victim's oxygen demand becomes too great and too many of his hemoglobin molecules surrender their cargo of oxygen—as can happen under physical stress—the oxygenless hemoglobins change shape in a subtle but abnormal way. The new shape, dictated by the laws of chemistry and physics acting within the hemoglobin molecule, predisposes them to join together, to *polymerize.* The defective hemoglobin molecules have front and back ends that happen to be complementary shapes, like cans from the grocery that stack stably in a certain way because the bottom of one fits snugly into the top of another. When the hemoglobins stack, however, they stick. Like railroad boxcars with a mind of their own, they quickly link into long chains, or polymers.

Spontaneously locking head to tail, the polymers keep growing as long as there are more abnormal hemoglobins to join. Then several chains twine around one another, forming a rigid cable. Like sticks somehow lengthening inside balloons and pushing out the elastic walls, the runaway hemoglobin polymers stretch the red blood cell's plasma membrane into oddly shaped forms—typically long, spindly sickle shapes—that clog the bloodstream. As tissues of the body become starved for oxygen, more abnormal hemoglobin molecules lose the oxygen that ordinarily keeps them from polymerizing. More misshapen red blood cells clog the blood vessels, enlarging the painful region of blockaded blood flow.

Most crises, as these episodes are known, eventually abate on their own because the sickled cells commit suicide. The logjam breaks, and blood flow resumes or improves, bringing new oxygen. Once in a while a crisis damages vital organs before it can abate, and the victim dies.

Like sickle-cell anemia, about 4,200 other human diseases are caused by a defect in a single gene that leads cells to make misshapen proteins that cannot do their job or to make no protein at all. In many cases, researchers know nothing about the gene involved except that it's flawed. In others, not even this is known. They deduce from the biochemistry of the disease and from the pattern in which the disease is transmitted from parent to child that a single gene is involved (one particular substance is missing or nonfunctional). For example, since most genes exist in at least two copies in each cell (one contributed by each parent), the disease usually does not appear unless both copies are defective. This is because the genes work independently, and in most cases either one alone can meet the cell's needs.

In recent years geneticists have gone much further. They have been able to track down the genes responsible for many diseases, learning how to detect their presence in a cell even if their malign effect has not yet been felt. Geneticists have found the genes whose defects lead to muscular dystrophy, cystic fibrosis, Huntington's disease, Tay-Sachs disease, Gaucher's disease, and a number of others. They can even read the gene's code and find the error. In other diseases, researchers have isolated not the gene itself but markers (which are other genetic sequences located close to the gene) that signal the presence of the disease. These include hemophilia, neurofibromatosis (or "Elephant Man" disease), and a nerve disease called amyotrophic lateral sclerosis (ALS), which was once better known as Lou Gherig's disease but is now probably more widely recognized as the disease of Stephen Hawking, the British astrophysicist.

Sickle-cell anemia, which afflicts three in every 2,000 black children and a smaller share of whites (among whom the incidence is highest in those of Spanish, Italian, Greek, Turkish, and Indian ancestry), is perhaps the most intimately understood genetic disease. Studies of sickle-cell anemia—tracing the agonies of a child to a single misplaced molecule in the cells of his or her parents—have helped shape biology's understanding of how deeply life, in sickness and in health, depends on *genes.*

In the living-room cell, the genes reside inside that VW Beetle of a nucleus. They are in the form of forty-six chromosomes—forty-six incredibly long threads, each made of a single molecule ranging in length from three miles to eighteen miles (remember the living-room cell is one that is scaled up 300,000 times). Like cobwebs in a long-abandoned barn, the threads festoon the inner surfaces of the nucleus, attached at many places. Even in a cell of normal microscopic size, the chromosomes are startlingly long. If the DNA from human chromosomes were stretched straight, the forty-six life-sized strands would range from three-quarters of an inch to three inches in length. If linked end to end, they would reach about 6 feet. Imagine 400 yards of actual-length DNA packed into the 200 cells that would fit in the dot of this exclamation point!

Incidentally, readers who have seen pictures of chromosomes that look like fat Xs may have come to think that that's how chromosomes always look. Actually, the X is a form they take only during a brief time at cell division when each thread goes dormant and shrivels and coils into a tight bundle. In fact, each X is actually two identical chromosomes fastened together at the middles (Figure 5.1). In this fat, linked state they are about to separate, each to head into one of the two daughter cells. (More about this in Chapter 6.) Chromosomes do not have this shape during the working phase of their existence. In those times, chromosomes are exceedingly slender threads. Even at the scale of the VW nucleus, they are like sewing thread, a total of 340 miles of it running every which way (Figure 5.2).

The threads are made of what is perhaps the most famous chemical abbreviation in science—DNA. It stands for *deoxyribonucleic acid,* a term hardly anyone, even in molecular biology, ever needs to use. Like so many of life's molecules, DNA is a polymer, a long chain of similar subunits, linked in various lengths. DNA's subunits, or monomers, are called *nucleotides.* A gene, then, is a specific series of nucleotides, usually between a few hundred and a few thousand. The 861 letters of the hemoglobin gene, for example, are 861 nucleotides.

In the entire human genome there are so many nucleotides of genetic

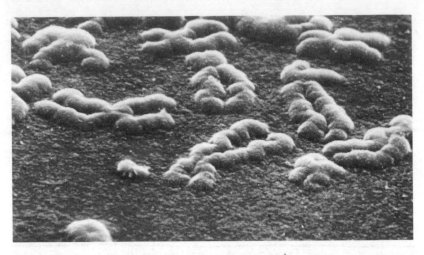

**Figure 5.1**  As seen here with a scanning electron microscope, human chromosomes are highly shriveled up. Their great length of DNA double helix has been coiled, supercoiled, and, for all practical purposes, wadded into this shape. DNA takes this shape only during cell division. Each of these three "fat X's," as they are called in the text, consists of two identical chromosomes joined at the middle. During cell division, the two will come apart and move into separate cells. (Photo courtesy of J. B. Rattner)

code that if they were printed as letters in a book, they would fill more than 500,000 pages. That's about 1,000 thick books. While this total can be estimated from the amount of DNA in a human cell, nobody knows exactly how many genes it includes. This is because genes vary enormously in length and because much of the DNA is known to be devoted not to genes per se but to regulatory sequences that determine which genes are active at any given time, spacers between genes, and other often unknown purposes. The usual estimates put the number of genes somewhere between 50,000 and 100,000. A few specialists say there may be as many as 250,000 genes. Whatever the true number, each gene carries, in the sequence of its linked nucleotide letters, the coded instruction for making one kind of protein molecule. The combination of proteins specified by the genes plus the structures formed by combinations of those proteins and the chemical reactions that other proteins can perform—all of this together—is what dictates the form and function of each cell. And, of course, the forms and functions of all the cells together dictate the form and function of the whole organism.

Genes are often presented to nonscientist audiences as the factors that determine eye color, skin color, height, and other fundamentally trivial attributes. They do, but the role of genes is far more profound. They also determine that a human being has two legs and can talk. They ensure that

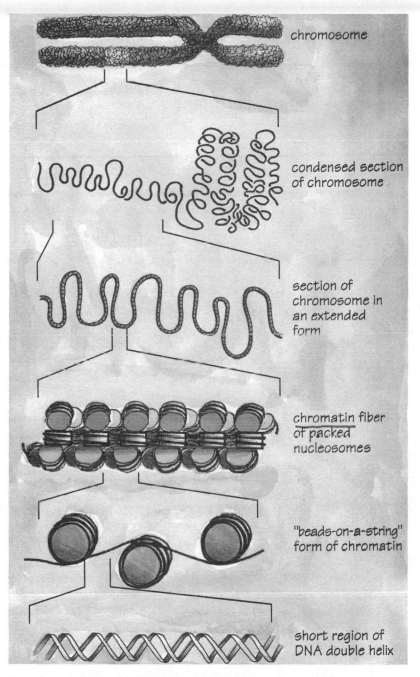

chromosome

condensed section
of chromosome

section of
chromosome in
an extended
form

chromatin fiber
of packed
nucleosomes

"beads-on-a-string"
form of chromatin

short region of
DNA double helix

**Figure 5.2**   The relationship between the double helix and the chromosome is shown in this series of artistic magnifications. Reading from the bottom, the double helix strand is looped twice around a series of protein molecules called histones. These "beads on a string" are then coiled to make chromatin fibers. These fibers are then coiled again and again to make the fully condensed chromosome.

the head sits atop the neck and that the heart pumps rhythmically. They give us an opposable thumb and a large brain. Genes also govern such obscure events as the extraction of energy from food, the way nerve cells store memories, and the way red blood cells deliver oxygen—or in the case of sickle-cell anemia, the way red cells sometimes fail to deliver oxygen. In other words, genes govern—or at least have a powerful say in—everything from the molecules in cells to the design and workings of the whole body.

For all their importance, genes are perhaps the most passive parts of the cell. They do not, so far as can be determined, perform any physical activity. They administer their realms with regal aloofness, always staying safely inside the nucleus, sequestered from the biochemical hurly-burly just outside. Like a computer program residing passively on a magnetic disk but controlling a vast, automated assembly line, genes simply let their messages be read. The messages—sentences of genetic code—direct the cell's manufacturing organelles to assemble amino acids into various kinds of protein molecules. There are twenty different kinds of amino acids and they can be linked in millions of theoretically possible sequences and lengths.

All human cells, with a few exceptions, (which we'll get to shortly) contain identical copies of the complete set of human genes. But relatively few of those genes—perhaps between 5,000 and 10,000, or between 5 percent and 20 percent—are active in any one cell. Most of the active genes are for enzymes that perform routine housekeeping chores, the metabolic processes fundamental to any unit of life, and for the structural proteins that make up the architecture common to all cells. Only a few genes, perhaps 100 in any given cell, contain codes for proteins needed only in that type of cell. For example, all cells contain the gene for insulin, the substance the body needs to metabolize blood sugar, but the insulin gene remains dormant throughout the body except in certain cells of the pancreas. Likewise, all cells contain hemoglobin genes, but they are switched off in all but the cells destined to become red blood cells.

This is one exception: All mature red blood cells in humans contain no nuclear genes. They don't even have nuclei. Red cells are formed from the division of other cells in the bone marrow, and for unknown reasons, their nuclei are pushed out of the cell and destroyed once they have issued their genetic instructions. But because the necessary commands have already been transcribed into another form, the cell can live and function perfectly well for months. Red cells normally live for 100 to 120 days before dying. As it happens, the red cells of people with sickle-cell anemia live only about half as long. This means they have fewer red cells per drop of blood, causing anemia. Another exception to the rule that all cells have nuclei is the *platelet,* a comparatively tiny blood cell, which is a piece pinched off from a much larger parent cell. Platelets help stop bleeding and will be discussed in Chapter 10.

Most genes exist in at least two copies, one from each parent. Some genes are present in a great many copies, half the total coming from each parent. People inherit some genes that are unique, only one working copy per cell. They are borne on either of the sex-determining chromosomes, which are designated X or Y. (Males have one X and one Y. Females have two X's, but one is permanently inactivated.) In sickle-cell, the child's prognosis depends on whether one or both parents passed on an abnormal gene. The cells of about 8 percent of African Americans in the United States carry one copy of the abnormal hemoglobin gene along with one normal version. These people are said to be "carriers" and to have "sickle-cell trait," a condition that almost never causes trouble in those individuals because they have one good hemoglobin gene, which is usually sufficient. Since both genes function, their red blood cells contain both normal and abnormal hemoglobins. The presence of the normal type usually keeps the abnormal from forming polymers. If two carriers have a child, however, there is a one-in-four chance that both of the child's hemoglobin genes will be abnormal and, thus, the child will have the disease. About 1 in 400 American blacks has sickle-cell anemia, many of whom die as children or young adults from a combination of the anemia and the effects of crises.

Apart from its significance as a cause of human suffering, sickle-cell anemia occupies a historic position in the rise of molecular biology. In 1956, sickle-cell's abnormal hemoglobin became the first protein whose altered function could be traced to a specific error in the sequence of amino acids making up its structure and thus to a specific mutation in the gene that governs the manufacture of hemoglobin.

Curiously, each human cell contains far more length of DNA than is needed to account for the presumed number of genes, perhaps ten or twenty times as much. Thus, of every twenty feet of DNA thread in the VW Beetle nucleus, only one or two feet is a true gene sequence that encodes a protein's structure. Some of the rest of the DNA plays regulatory roles, carrying codes that determine which genes are active when. Other parts of the DNA may have other roles. But most of the rest of the DNA is probably just plain junk—long stretches of nucleotidal gibberish.

There is an intriguing theory that some of the genetic junk consists of wrecked genes, codes that made sense and were active in our evolutionary ancestors but which, like antique muskets in museum cabinets, are permanently disabled with the biological equivalent of having their firing pins removed. If so, we may carry decommissioned genes for claws and fangs, remnants of ancient genes for scales to cover our skin, maybe even scraps of very old genes for fins. The ancient genes—or former genes,

since they don't work in humans—keep getting passed on from generation to generation along with the currently working genes (perhaps because cells lack the apparatus to eliminate them) but they are now only
filler, only so much spacing between the active genes of today. Indeed,
studies of junk DNA have shown that it contains nucleotide sequences
that resemble real genes but with odd garblings of the coded message.
They are sometimes called *pseudogenes*. Much of this genetic gibberish
may also play an important role in evolution. We'll get back to the deeper
significance of that speculation later in this chapter.

If you walk over to the VW nucleus in your living room and look
inside, you might just barely see a piece of the DNA thread with a
magnifying glass. If you look very closely, you might also make out one of
the best-known shapes in modern biology—the *double helix*. Its discovery
in 1953 led to Nobel prizes (1962) for James Watson, Francis Crick, and
Maurice Wilkins. It was their discovery, said Jacques Monod, the French
pioneer of molecular biology, that finally gave Darwin's theory of evolution its full significance, weight, and certainty.

A helix is the shape of a corkscrew. A double helix is the shape of two
corkscrews, one intertwined with the other and curving parallel to it, like
the railings of a spiral staircase. Another way to think of the double helix
is as a twisted rope ladder with rigid rungs, each rope forming a helix.
The real significance of Watson and Crick's discovery was not the helical
nature of DNA but its "doubleness." The key discovery, the finding that
opened the door to the research that has confirmed Darwin so dramatically and made possible the surging advance of modern genetic research,
is that the molecule of heredity is made of two parallel strands, that each
strand is made of a sequence of linked elements (nucleotides), and that the
parallel sequences are structurally complementary to one another. This
simple but profound insight instantly revealed how the genetic code could
be copied over and over again and passed so accurately from one generation to the next. The discovery also led quickly to an understanding
of how errors could be introduced on rare occasions, errors that would
be the mutations that make evolution possible and, unfortunately, to
diseases like sickle-cell anemia that are passed on from one generation to
the next.

The easiest way to think about DNA is to start by splitting the two
helices apart. Think of it as sawing down through the middles of
the wooden rungs of a rope ladder. The result is two single ropes with
half-rungs hanging off each rope (Figure 5.3). Only one rope—one strand
of the double helix—carries the genetic code. The other usually acts
simply as a "keeper," a complementary strand that fits against the coding
strand. (During cell division, subject of the next chapter, it plays a more
critical role.)

In all DNA the rope part of the ladder is the same up and down its

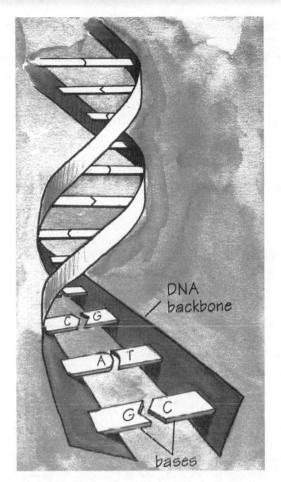

DNA
/ backbone

C G

A T

G C

bases

**Figure 5.3** A closer look at the double helix in Figure 5.2, this shows the all-important "bases" that project from each DNA strand and that link to form rungs in the ladder. There are four kinds of bases, abbreviated A, T, G, and C. Because of the shape of each base, an A will link only with a T and a G only with a C. This is the principle that makes each strand the chemical complement of the other. The principle of complementarity is crucial to a later step in which a strand of messenger RNA will be made as a complement of the DNA strand. It will also come into play again during DNA replication, the subject of Chapter 6.

length—a monotonous alternation of sugar and phosphate molecules. The half-rungs, which stick out from the sugars, are the interesting parts. They serve as the four letters of the genetic alphabet—four different kinds of molecules abbreviated A, T, G, and C for adenine, thymine, guanine, and cytosine. These four molecules are called *bases*. (The nucleotide is a more inclusive unit consisting of one phosphate, one sugar, and the base sticking out from the sugar.)

As in the English language, the sequence of bases along one strand of DNA—the sequence of As, Ts, Gs, and Cs—spells out the genetic message. A sentence in genetic code—a gene—may start out ATCGCGAAT and continue in that vein for hundreds or even thousands of letters, or bases. On the opposite strand would be the complementary sequence. Opposite every T would be an A (and vice versa) and opposite every G would be a C (and vice versa). Thus the complementary sequence for the one just given would be TAGCGCTTA. Complementarity is a result of the shapes of the bases. Only an A will fit a T; only a C will fit a G.

The gene for hemoglobin, as it happens, is actually two genes because hemoglobin is made of two slightly different proteins, designated alpha and beta. Each hemoglobin is a modular protein, containing two alpha modules and two beta modules, which fit together in a four-leaf-clover pattern. The code for the alpha protein has 423 letters, or bases; the beta protein code has 438. Together they make 861 bases.

The difference between a person with normal hemoglobin and one with sickle-cell anemia—the difference between a normal childhood and one marked by painful and life-threatening crises—is the seventeenth base in the beta protein's code. The normal code has a T; the abnormal, an A. Because cells dutifully obey the instructions of their genes, they respond to this mutated message by placing the wrong amino acid at a certain position within the protein being made under the guidance of the gene. The change from a T to an A alters the meaning of the genetic "word" containing the letter. Instead of the amino acid called glutamic acid, the abnormal gene instructs the cell to use one called valine. Valine fits into the sequence (amino acids can be linked in any sequence) but it distorts the shape of the finished protein, rendering it less able to function as an oxygen carrier. Valine and glutamic acid are two of the twenty different amino acids that can be linked to make thousands of different protein molecules depending on the sequence in which they are put together.

The four letters of the genetic language—the four kinds of bases—are formed into various three-letter words, each of which stands for one of the twenty kinds of amino acids. Molecular biologists call each word a codon. Actually, there are sixty-four possible three-letter (three-base) combinations. In many cases, several genetic words are synonyms that specify the same amino acid. Three codons function as "stop" signs marking the ends of genetic sentences. (There are also codes that say "go" at the beginning of each gene.)

The mutation that causes sickle-cell anemia lies in the sixth codon of the beta hemoglobin gene. It is supposed to read "CTC," telling the cell's protein-making machinery to install glutamic acid as the sixth amino acid in the chain that will become the beta protein. The abnormal mutation, "CAC," tells the cell to put valine in that position.

One of the profound discoveries about the genetic code is that all living cells use essentially the same code. There are relatively minor exceptions in which the code for certain amino acids in some organisms is different, but otherwise this discovery dramatically confirms the Darwinian view of evolution. It means that all life on Earth must have descended from one common ancestor. It does not mean that other forms did not arise but it does mean that if they did, they died out and left no known survivors.

The reason this must be so is that there is no physical reason why a given DNA codon should specify, or "mean," any particular amino acid. The meaning of any given codon is determined by a double-ended molecule that recognizes a specific genetic sequence on one end and carries a specific amino acid on the other. It's called transfer RNA, and we'll discuss it a little later. This molecule could just as easily link the same codon to a different amino acid. Think of the way libraries use code numbers in their catalogues to identify specific books. The link between the number in the card catalogue and the book is arbitrary. Obviously two libraries may use completely different numbering systems to indicate the same title, and each could find its own books perfectly well. But, if all the libraries in the world used the same catalogue numbering system, it would be powerful evidence that they had copied their systems from one another. In the same way, it is inconceivable that all the living species on Earth could have accidentally hit upon the same arbitrary system that links triplet codons with specific amino acids.

Many other genetic codes are equally conceivable—just as there are, in fact, several different library cataloguing systems and many different human languages with different words that have the same meanings. It is well known that spoken languages evolve so that if one group of speakers becomes isolated from another long enough, the accumulated changes in the two languages will eventually produce two different languages. Italian and Spanish, for example, are both descendants of the Latin that the ancient Romans spread across Europe. If biologists found widely differing genetic codes among species living today, they could legitimately wonder whether they evolved from a common origin or began separately. As it happens, the genetic codes are everywhere identical, and the exceptions are still so close that they are like distinguishing American English from British English.

Because the genetic code is universal, genes from one species have meaning when implanted in the cells of another species. This fact is the basis of the biotechnology industry, which has put, for example, the human gene for insulin in bacteria. The bacteria have no trouble reading the human gene and making human insulin. Bacteria are now a source of human insulin, which is extracted and given to people with diabetes. In the past, diabetics had to use insulin extracted from pigs, which was slightly different from human insulin and sometimes caused allergic reactions.

For all their power, genes do not act autonomously. Most exercise their supposed dictatorial role only when switched on by other molecules in the cell—a fact that leads to one of the subtlest but most important points in all of biology. Cell and molecular biologists are proving that human beings are not the products solely of their genes. They are also the products of their environments. For it is the environment—both that of the individual cell and that of the whole person—that determines which genes are switched on and when. Some of the molecules that turn genes on, and turn them off, are produced within the same cell. Some, such as certain hormones, come from cells in other parts of the body, slip right through the cell membrane, and head straight into the nucleus to act directly on the DNA. Others from elsewhere in the body act on receptors in the cell membrane, which relay a signal to the DNA. Even environmental influences from a person's surroundings can turn genes on and off. Stress from frightening or difficult circumstances, for example, causes cells in some glands to start manufacturing certain hormones. Mental activity of various kinds can cause brain cells to send signals and sprout new tentacles that make richer contacts with other cells. The act of learning can regulate genes in brain cells, creating more powerful and effective minds.

Northwestern University's Robert D. Goldman has speculated that environmental influences may act via cell surface receptors that are directly linked to the DNA—and perhaps even to specific genes—by networks of fibers that are part of the cytoskeleton. Though once thought to be a passive structural component, the cytoskeleton is now known to consist of several very different networks of filaments lacing their way throughout the cell. Goldman suspects that the medium-weight fibers, called *intermediate filaments,* may form direct links between receptors and genes. He has shown that elaborate networks of intermediate filaments link various parts of the nucleus to thousands of points on the cell surface. A signal from an activated receptor might travel along the filament carried by an as-yet-unknown motor molecule or the signal might be relayed by an intrinsic change in the filament itself. Intermediate filaments, like all others, are polymers of small subunits. A force that altered the monomer at the receptor end might have a kind of domino effect, changing the shape or orientation of the adjacent monomer, and so on straight back to the DNA, where a change in the last monomer might somehow activate the gene. Such a scenario is highly speculative now but it illustrates the kind of mechanistic thinking that is possible given today's deeper understanding of the fundamental workings of life.

Much less speculative are other parts of the gene-regulation story. LDL, the cholesterol-carrying protein described in the previous chapter, is a good example. The activity of the LDL receptor gene is regulated by the amount of cholesterol inside the cell. In other words, when the cell is low in cholesterol, the gene's accelerator is floored. The gene's message causes more LDL receptors to be manufactured and installed in the cell membrane. This improves the cell's ability to pull in LDL from the bloodstream, along with its attached cholesterols. But as the internal concentration of cholesterol goes up, molecular messages reach the nucleus and apply the brakes to the LDL receptor gene, slowing or stopping the manufacture of new LDL receptors. The system is rather like a thermostat, turning up the receptor gene's activity when the cell needs cholesterol, turning it down when it has enough. If cells can produce enough LDL receptors to get all the cholesterol they need, they are happy. If not, the cells can make their own cholesterol. This is one of the reasons that many people cannot totally overcome their high-cholesterol problem simply by cutting down on the amount they eat. When people eat less cholesterol, their cells simply make more. Since cholesterol is not a protein, there is no gene for it. Cells make it through chemical reactions controlled by protein enzymes for which there are genes.

The quantity and activity of gene-regulating molecules often depends on the environment surrounding the whole individual. Along with stress or learning, other environmental factors also influence gene activity. Gene-regulating substances can come from the diet, from viruses infecting the body, and, perhaps most often, from chemical messages sent out by other cells in the body.

Each cell's nucleus, then, is at the center of a veritable symphony of chemical signals arriving from all directions. Its behavior depends on the combination of all the outside influences. There may, for example, be gene-activating molecules arriving at the same time as molecules that would inhibit the same gene. The two may interact or compete in other ways to influence the cell one way or the other. At the same time, each cell may be giving off chemical messages that affect others throughout the body. The life of a multicelled organism depends on an extraordinarily complex interplay of thousands of different chemical signals coming from each cell and going to some or all of the other cells that make up the republic of cells that is the human body.

Whatever the source of the signal, its effect is to work within the nucleus, seeking out the regulatory portions of DNA that govern the specific gene it "wants" to activate. The signal proteins must find and bind to the regulatory segments to cause a given gene's message to be "expressed" or, depending on the signal, to prevent expression. The sequence of DNA bases that constitute a gene's "on" switch is positioned next to the "go" signal at the beginning of the genetic sentence. Also taking part in the process are other regulatory sequences that are often situated some distance away from the gene. At one time it was assumed that the signal proteins wandered aimlessly within the nucleus until they chanced to bump into their target sequence. Now it is clear that they don't. They grab onto a DNA strand and "walk" along it, "looking" for the sequence—yet another example of the role of autonomous motion in life. Even though the molecule may have to walk a long way to find its gene (many miles in the living-room cell nucleus), the process is roughly 100 million times more efficient than simply bouncing around inside the nucleus.

The way all the genes, regulatory sequences, and signal proteins work together is extraordinarily complex. Fortunately, molecular biologists had a clue from their work on the much simpler way that bacteria regulate their genes. We'll start there.

A typical bacterial gene has an adjacent sequence called the *promoter region*. This is a sequence recognized by one of the largest and most versatile molecules in all of biology. It's called *RNA polymerase*. This is the enzyme that carries out the first step in gene activation—making a copy of the gene's code. If RNA polymerase can bind to the promoter region, it will unzip the double helix, separate the two strands, read the gene's code, and begin synthesizing a complementary strand of a similar molecule called *messenger RNA* or *mRNA*. This form of RNA is the molecule that will carry the gene's code out of the nucleus and into the protein factory elsewhere in the cell.

Sometimes, however, RNA polymerase cannot bind to the promoter region because there is another molecule in the way. A *repressor protein* may be stuck to a region called the operator within the promoter. As long as the repressor protein is there, the gene's code remains unread. So to switch the gene into action, the signal molecule must bind to the repressor, causing it to change shape and fall away from the regulatory region. Once that happens, RNA polymerase can grab ahold and start transcribing the gene (DNA) into messenger RNA.

In some cases the signaling works the opposite way. For example, bacteria have a gene for an enzyme needed to manufacture the amino acid called tryptophan. As long as the cell has enough tryptophan, some of it binds to the repressor, keeping it in the form that stays stuck to the operator. This is efficient because the bacterial cell has tryptophan and doesn't need to make more. But if the number of tryptophan molecules falls, there are not enough to stay bound to the repressor. As a result, the repressor falls away from the operator, exposing the promoter region so that RNA polymerase can grab on and start the process of making more tryptophan (Figure 5.4).

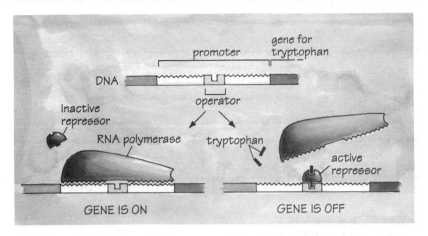

**Figure 5.4** Genes are switched on and off by various ingenious sets of interactions among regulatory segments of DNA near the gene and other molecules that bind to DNA or otherwise interact with it. Here, at the top, is the gene that guides the making of tryptophan (only the beginning shown) with a "promoter" region next to it. If the RNA polymerase molecule can grasp the promoter, it will proceed to read the gene and transcribe its code into the form of messenger RNA. This is shown at the bottom left. When it is not needed, the gene can be turned off simply by the presence of tryptophan. This is because tryptophan binds to a repressor molecule, changing the repressor's shape so that it, in turn, binds to the "operator" segment of the promoter. With the repressor in the way, RNA polymerase cannot get to the promoter to make more messenger RNA.

In the cells of more complex organisms, including humans, DNA regulation is similar but more complicated. Instead of just one activator or repressor for a given gene, there are commonly several, sometimes a dozen or more. Some bind to the promoter region of the DNA, an action that changes their shape. Then another protein binds to the changed shape of the first and changes shape itself. A long series of similar bindings and shape changes may be required before the final agglomeration of proteins attains a shape to which RNA polymerase will attach and start transcribing the gene.

Why do we have such a complicated process just to start reading a gene? Mainly because it allows for much more sensitive and variable regulation of the rate at which genes are activated. Genes don't just express their messages at any old time. They do so only when and where the gene's product is needed. In a bacterium, this is a fairly simple decision. Only one cell is involved, so one "on" or "off" signal seems to be enough. But human cells must function in harmony with trillions of others. Each one of the regulatory proteins is needed, but the abundance or scarcity of any one of them can influence the number of times that the right combination comes together to make a gene go.

While some regulatory proteins encourage gene activation, others repress it. Molecular biologists believe that in some cases, proteins of both kinds compete to affect a gene. Some DNA-binding proteins stay stuck on the DNA for a long time, while others bind only weakly and may easily pop off and float away. Varying levels of competitiveness—repressors battling activators—may result in levels of gene expression that vary by different amounts. In other words, competition between the two regulators could make it so that a gene can be expressed repeatedly many times in a short period of time or only a few times or never. Subtle changes in the relative amounts of molecules that bind to DNA, or to other DNA-binding molecules, with different strengths could produce an exquisitely sensitive regulation of gene activity.

Amazingly enough, some of the regulatory DNA sequences are not situated near the gene they control. They may be hundreds or tens of thousands of bases away. And yet, if the right regulatory proteins don't bind to them, the gene won't be expressed. How can this be? How can the gene or the RNA polymerase "know" whether certain proteins have grabbed onto the DNA strand a long distance away? Simple. DNA is flexible. It simply forms a loop in itself, bringing the distant regulatory protein into contact with other proteins on the promoter. In the case of the beta hemoglobin gene, there are several regulators—several that speed up the gene-activation rate to varying degrees and some that slow it down to varying degrees. Somehow they work it out among themselves how often the gene should be allowed to express itself (Figure 5.5).

Whatever happens, the result—if the activators prevail—is to unleash RNA polymerase, which makes messenger RNA. (In the lingo of biochemistry, incidentally, the suffix "ase" on a chemical's name indicates that it is an enzyme that performs some function related to the first part of the name. Polymerase, in other words, means an enzyme that makes a polymer. Unfortunately, the naming system is not consistent. The "ase" can also mean it is an enzyme that breaks down the chemical designated in the first part of the word. A protease, for example, is an enzyme that breaks down proteins.)

Just like its close chemical cousin DNA, messenger RNA (for ribonucleic acid) is a long-chain polymer of nucleotides (in RNA they are called *ribonucleotides* because the sugar in the sugar–phosphate backbone is a form called ribose). Like DNA, RNA contains bases whose sequence encodes a genetic message. But the code is slightly different. A strand of messenger RNA is a transcription of a DNA gene. The relationship of DNA and messenger RNA are like that of a computer program permanently stored on a disk (DNA) and the temporary copy of the program loaded into the computer's active memory (messenger RNA). In other words, it is a temporary copy made for use during a brief interval and will soon be discarded or broken down.

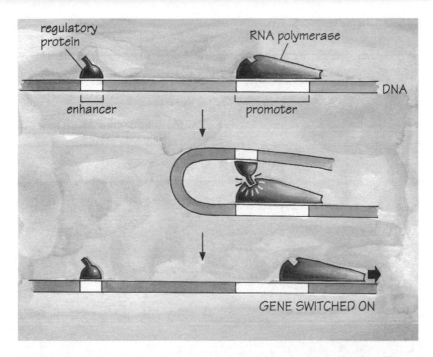

**regulatory protein** ... **RNA polymerase** ... **DNA** ... **enhancer** ... **promoter** ... **GENE SWITCHED ON**

**Figure 5.5** Curiously, regulatory segments of DNA do not need to be adjacent to the gene they regulate. Here is a schematic of how such a mechanism can work. At top, the RNA polymerase has found a promoter region but still needs to be activated. Some distance upstream of the DNA strand is an "enhancer" region to which a regulatory protein binds, changing its shape ever so slightly. Because DNA is flexible, the strand is free to bend (middle) so that the regulatory protein can interact with RNA polymerase. This activates the RNA polymerase.

While the permanent copy of the gene stays safely in the nucleus in the form of DNA, messenger RNA carries a copy of the instructions out to the cell's factories, to the cell's protein-making machinery. In another analogy, messenger RNA is the foreman who has read the blueprints in the office and goes out onto the factory floor to direct the making of the product according to the specifications.

RNA polymerase, the enzyme that makes messenger RNA, can do its job only after the regulatory proteins have docked with the gene's regulatory region. In other words, the regulatory proteins pick the gene to switch on and the RNA polymerase obligingly transcribes whatever gene the regulatory proteins have selected.

It does this by stepping down the length of the gene like an inchworm on a twig, unzipping the two strands as it goes and exposing the one coding strand of DNA that carries the code. If you put a magnifying glass to the threads festooning the inside of the VW nucleus, you could see many huge RNA polymerases walking down lengths of DNA, one base at

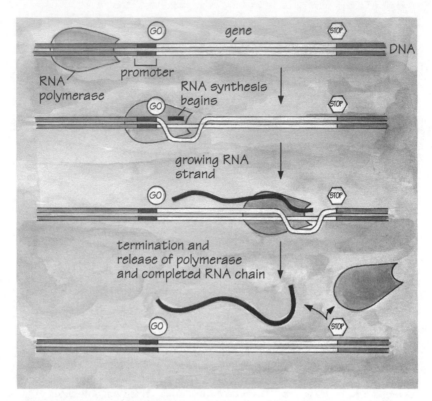

**Figure 5.6** Once activated, the RNA polymerase molecule works its way along the gene, making a strand of RNA in which each RNA base is the complement of each DNA base. In this case, the RNA complement is slightly different from that of DNA. Where the complementary DNA base would be a T, RNA has a different base, abbreviated U. Once RNA polymerase reaches a DNA code that says "stop," it falls away from the DNA strand and the strand of messenger RNA is released.

a step (Figure 5.6). At each step, the polymerase grabs one of the free-floating RNA nucleotides, each bearing one of four bases, and links it to a growing sequence that is the mirror image, or the complement, of the DNA sequence.

In both DNA and RNA, to reiterate, complementarity means that if one strand has a certain base at a given position, the corresponding strand will have the complementary base at the same position. In DNA there are just two complementary pairs—A plus T and G plus C.

In an RNA transcription of DNA, the complements are the same except that where the DNA complement would put a T, RNA puts a U (for the base called uracil).

As the RNA polymerase ratchets down the gene, it pushes out—segment by segment, like a craft worker stringing beads—a newly minted strand of messenger RNA. Its sequence of bases corresponds to the sequence in the DNA strand. With each step, the RNA polymerase adds a new ribonucleotide to the one it added just before.

When the RNA polymerase comes to the end of the gene, it encounters a special sequence that stops it from moving any farther. The enzyme falls off the gene and releases the messenger RNA chain.

At this point an extraordinary thing happens, something that molecular biologists considered one of the greatest surprises of modern genetics when it was discovered in the 1970s and early 1980s. They found that genes are not straightforward sequences of meaningful bases. They are interrupted by long stretches of what appeared to be gibberish, lengths of DNA in which the nucleotide sequences are garbled and nonsensical. It was as if, in reading an English sentence qwxtrjupodnbeytrh, you came across nonsense sequences of English letters that idngketfrzl interrupted the real message. No one knows for sure how these intervening sequences, now called *introns,* arose. But there is growing evidence that introns play an important role in the evolution of new genes and, hence, new proteins. We'll examine the evidence later in this chapter. Subsequently, researchers have found that the sequences within introns are not entirely meaningless; some bear codings that help regulate the function of the gene.

The newly made messenger RNA contains a faithful transcription of the gene, including the introns. So, before the messenger RNA can go to work, these sequences of genetic gibberish (plus any regulatory segments within them) must be removed. Again, special enzymes do the job. There are, in effect, editor enzymes that can detect a special code—AGGU—at the beginning of each intron and AGG at the end. The editor enzymes cut the messenger RNA chain between the two Gs at each end of the unwanted segment, throw away the gibberish and splice together the loose ends of the good segments, which are usually called *exons* because they are the sequences that are expressed (Figure 5.7).

Once the nonsense has been edited out, the messenger RNA emerges from the nucleus and travels to the cell's protein-making machinery. Exactly how this happens nobody knows. Unlike a VW Beetle, a nucleus has no windows or doors. It has only tiny pores, no bigger on the living-room scale than nail holes. Presumably the messengers emerge through these pores. And it is also presumed that there is some active method of transporting them, some courier. Perhaps a yet-undiscovered motor molecule does the job. Perhaps some entirely unknown mechanism is at work.

In any event, the thread of messenger RNA moves out of the VW nucleus and travels until it is embraced by one of the smallest of the cell's organelles, the *ribosome.* Hundreds of thousands of marble-sized ribosomes float free in the living-room cell, drifting among and around the other objects.

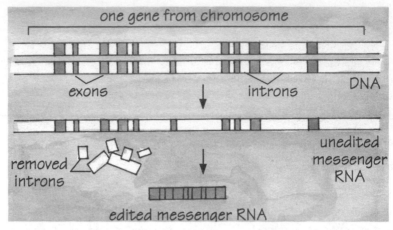

**Figure 5.7** Most genes contain intervening segments of gibberish, stretches of DNA bases that are not supposed to encode anything in the finished protein. But RNA polymerase cannot detect them and faithfully copies these gibberish segments into the messenger RNA strand. They must be removed before the messenger RNA is used to guide the making of proteins. Special "editor" enzymes travel the length of the unedited messenger RNA, removing the gibberish (white segments, called introns) and splicing together the remaining useful segments (gray, called exons).

Each ribosome is a complex machine—made up of some thirty kinds of proteins and three kinds of nucleic acids like those in RNA. If the messenger RNA is the foreman bearing a working copy from the master plans in the office, the ribosome is a factory worker who must read the working copy of the plans and assemble the specified product by fastening its components together. In the case of proteins, the components are all smaller molecules called amino acids. For example, the alpha chain of hemoglobin—the one that is okay even in people with sickle-cell anemia—contains 141 amino acids, each specified by one triplet codon in the messenger RNA. The beta chain, the one that can be defective, contains 146 amino acids. It is, to repeat, the exact sequence of the different amino acids that gives the protein its functional properties.

The ribosome automatically embraces a strand of messenger RNA, its two subunits grasping the RNA between them (Figure 5.8). It holds the RNA in such a way that codons are brought, one at a time, into a special position near a key part of the ribosome. Drifting loose in the watery liquid nearby is a stockpile of free amino acids, each held by a short piece of another kind of RNA called *transfer RNA,* or *tRNA.* Each transfer RNA molecule has two business ends. One is a single triplet codon of its own, and the other holds an attached amino acid. Each kind of amino acid

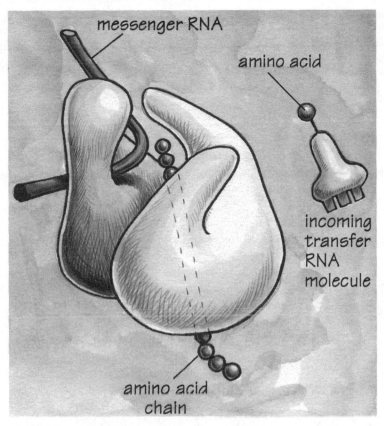

messenger RNA

amino acid

incoming
transfer
RNA
molecule

amino acid
chain

**Figure 5.8** A ribosome, made of two interacting proteins, grasps a strand of messenger RNA and follows its code (a transcription of the DNA code) to make a chain of amino acids, the building blocks of proteins. Each loose amino acid is carried by a molecule called "transfer RNA." The process is shown in greater detail in Figure 5.9.

is linked to its own specific triplet codon, a different codon for each kind of amino acid.

The ribosome holds a triplet codon of the messenger RNA in just the right position to bind to the triplet codon of any transfer RNA that happens to have the complementary code. For example, if the messenger RNA codon being exposed at the moment has the shape that says GAG, the complementary codon would have to read CUC. That is the only codon that will bind complementarily with the messenger's codon. And, because of the way a transfer RNA with the CUC codon is built, it will carry only the amino acid called glutamic acid (Figure 5.9).

The ribosome holds the glutamic acid in place, breaks off and discards

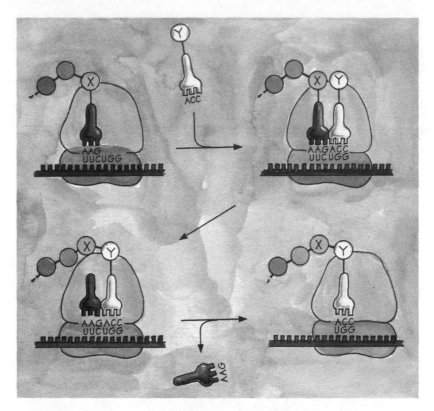

**Figure 5.9** Starting at the upper left, we see the blobby shape of the ribosome behind a strand of messenger RNA, each tooth representing one base. Look at the first three bases—UUC. Just above them is a transfer RNA bearing the code AAG. This is the complement of UUC and, so, binds briefly, holding its amino acid (labeled X), which is linked to a previously assembled chain of amino acids. Moving to the upper right, another transfer RNA arrives, bearing the code ACC and the amino acid Y. Because ACC is the complement of the next triplet of messenger RNA bases (UGG), this new transfer RNA binds and its amino acid (Y) is linked to the X. Next, at the lower left, the previous transfer RNA releases its hold on X and falls away. At the lower right, the ribosome has moved one step along the messenger RNA, bringing another set of bases (unlabeled) into position for yet another transfer RNA.

the transfer RNA that held it, and moves one step along the messenger RNA to bring its next codon into reading position. When another transfer RNA has locked onto this next codon, the ribosome brings that new amino acid into position to bind to the previous one. Again, it casts away the transfer RNA and moves on to the third codon.

Step by step, codon by codon, the ribosome grinds and cranks its way along the strand of messenger RNA like some marvelous contraption of levers and gears.

For all its complexity, the making of proteins is astonishingly fast. Moreover, many ribosomes may work their way along a single messenger RNA simultaneously, all churning out identical chains of amino acids that, when finished, will be protein molecules. In a typical single human cell, thousands of ribosomes are carrying out more than one million amino acid–binding reactions every second. In a single cell they manufacture an estimated 2,000 new protein molecules per second. The late Lewis Thomas, the former head of the Sloan-Kettering Memorial Cancer Research Center in New York and author of *The Lives of a Cell,* once said he liked to think he could feel his entire body tingling as billions of little ribosomes chattered away, clicking and whirring, spinning out new proteins.

It is in this process that the genetic error of a person with sickle-cell anemia is obeyed, a step that will doom the victim. The messenger RNA codon that should read GAG instead says GUG. This means that the only transfer RNA that the ribosome will accept in that position is one carrying valine. A proper hemoglobin has glutamic acid in that position. But no cell can ignore a genetic command. Into the nascent protein the ribosome installs the wrong amino acid.

There is more to a protein, however, than a chain of amino acids. Once the protein emerges from the ribosome, it folds and curls—vaguely like a potato skin from the peeler—into a shape determined by the chemical propensity of each amino acid to interact with others in the chain. Protein folding, as this phenomenon is called, is largely automatic, dictated by the chemical proclivities of the amino acids. It is also usually consistent; a given amino acid sequence will tend to fold up into the same shape as long as the chemical environment is the same (Figure 5.10).

Parts of the chain's length coil into helices, other segments lock into rigid rods, still others act as flexible swivels. One amino acid, called cysteine, binds strongly to other cysteines, forming virtual "spot welds" where the chain attaches to itself. Very long amino acid chains often fold into huge wads. Parts of chains loop out, double back, cut over to the opposite side of the wad, shoot out in a great helix, then return in a straight line to the center of the wad. Interaction with water in the surrounding fluid also plays a role. Some amino acids are hydrophobic, repelling water, and others are hydrophilic, readily soaking up the wet stuff. In many proteins there are sequences of hydrophobic amino acids that are pushed deep inside the wad, away from the surrounding water. It is usually the wad's outer contours and the chemical nature of the amino acids that wind up as part of the surface that give the finished protein its powers. Sometimes, however, the hardest-working part of the protein may be at the bottom of a deep pocket or cave. One enzyme, for example, is known to have a tunnel into which it pulls the substance it is to break down. Electrical charges on the outside of the tunnel attract the opposite

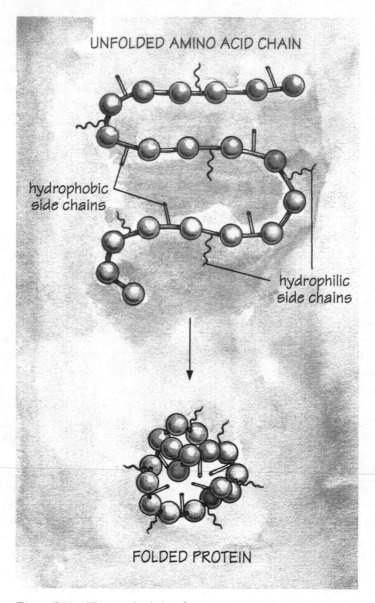

UNFOLDED AMINO ACID CHAIN

hydrophobic side chains

hydrophilic side chains

FOLDED PROTEIN

**Figure 5.10** The simple chain of amino acids produced by the ribosomes is far from a working protein molecule. It must first fold into a specific shape. Because each amino acid may have chemical attractions or repulsions with respect to others in its chain, these forces push and pull the chain into a characteristic wad. Some repeating sets of amino acids cause the chain to form a spiral. Others make the chain zigzag repeatedly.

charge on the target molecule, essentially sucking the hapless target into the tunnel. The molecule is broken apart and the pieces are ejected through another tunnel, again propelled by electrostatic charges.

Though much of the folding that can create such an elaborate machine is dictated by chemical forces within the amino acid chain, there may be several alternative folding patterns that a given sequence of amino acids can achieve. But only one is right. To help the chain find the right arrangement, cells have many kinds of *chaperone* molecules. These are proteins that grasp the new, partially folded chain and push or pull it into one particular shape. Like a sculptor modifying an unsatisfactory clay figure, the chaperones massage the amino acid chain, nudging a helix sideways or shifting a loop from this side to that. The name of chaperone, however, arose from a misunderstanding of their role. When cell biologists first discovered these molecules, they thought their main job was to keep different newly formed amino acid chains from coupling promiscuously before they could take on their final folded form. "Chaperone" should now probably be changed to "masseur."

With or without chaperones, the shape of a protein molecule can be exquisitely sensitive to changes in the amino acid sequence. Change it even a little bit in the right place and you may change the shape and functional ability of the whole, folded bundle significantly. Or you might not change it much at all. Some amino acids play more crucial roles in a protein than others. A molecule that might have acted as an enzyme to make a certain reaction happen could lose that ability because it no longer has the right shape. Or it might acquire some new power. The changes may be profound or they may be inconsequential. Sometimes the altered DNA sequences lead to radical changes such as might happen, say, if a car's tires suddenly changed shape and became square. Others might have no more effect than, say, changing the shape of a car's door handle. Just such a phenomenon, changing amino acid sequences, countless times over the eons of evolutionary time, is directly responsible for the existence of the great diversity of living species. This is the phenomenon at the heart of evolution.

Darwin could not have known it. The central event that makes evolution possible is an alteration in the coded message carried by DNA—in other words, a mutation. This change is ultimately translated by ribosomes into a protein of a new shape. And along with the new shape may come new powers to carry out chemical reactions or new proteins that alter the shape of the whole cell. Even the whole organism can be changed through mutations in genes governing the powers of cells to influence embryonic development. Change the specializations of cells and the sizes to which various organs grow, and you make the difference between a worm and an amphibian, a reptile and a mammal, an ape and a human being. It is a fact of biology that all these evolutionary differences arise entirely from a long accumulation of differences in messages coded in the genes and the proteins that are made in response to those messages.

Of course, the mutations that fall more into the realm of everyday concerns are the ones that arise within one's own generation to cause cancer, birth defects, and other maladies. These mutations can be caused by errors in the DNA-copying process during cell division (Chapter 6 is all about this) or by outside forces such as sunlight, radioactivity, and certain toxic chemicals. Ironically, these same causes also give rise to the mutations that make evolution possible. The chief difference is in the site of the cell being mutated. If the mutated DNA is in the gonads, where eggs and sperm are being made, the mutation can affect the next generation, for better or for worse. If the DNA is in ordinary body cells, the next generation will not be affected, but tumors could arise.

Wherever the mutations occur, they have no effect on the organism unless and until they are transcribed into messenger RNA and until that message is read by ribosomes to make proteins of new shapes and properties.

Obviously, over the billions of years of evolution on Earth, many mutations have accumulated, turning one or a few primordial proteins into thousands of different kinds. And yet, for all the diversity, there are many common themes underlying the structures of the different proteins. A great many proteins, scientists have learned in recent years, are not wholly novel structures. Instead, they are made of pieces copied from other proteins—not just arbitrarily chosen pieces but discrete functional units, modules of a molecule that play similar roles (for example, anchoring a receptor in a cell membrane or binding to DNA) in widely different proteins.

This, biologists have generally come to agree, is the reason for the surprising existence of those intervening sequences of genetic gibberish— the introns that have now been found in most of the genes that have been sequenced from organisms ranging from yeast to humans. They are spacers between exons, the segments of genes that are complete functional modules. The number of introns per gene can range from one to more than fifty. Introns can also be very long, sometimes much longer than the exons they adjoin. In fact it is estimated that somewhere between 75 percent and 90 percent of the average gene consists of intronic gibberish.

The receptor for LDL, the cholesterol carrier, for example, has eighteen exons (coding sequences) separated by seventeen introns (noncoding sequences). Some exons contribute the modules of the receptor that make up the socket into which LDL fits. One exon provides an attachment site where sugar molecules stick onto the protein. Two other exons supply the code for the hydrophobic module that anchors the protein in the cell membrane (it works because the inner part of the membrane is also hydrophobic). Yet another exon is for the tail that dangles inside the cell that interacts with clathrin, the geodesic-dome protein that helps the cell take in the LDL with its attached cholesterol.

Molecular biologists have also found that many of the exons in any one gene can be surprisingly similar to exons of seemingly unrelated proteins or, indeed, similar to the whole of other small proteins. In the LDL receptor gene, for example, there are exons that have very nearly the same genetic code as is in the gene for a protein called epidermal growth factor,which makes skin cells proliferate. The LDL receptor also has other exons closely resembling three different blood-clotting factors. And there is an exon much like the sequence for a protein called *complement* that plays a role in the immune system.

Similar instances of unexpected similarities among genes with very different functions have turned up again and again. So have discoveries that the exact positions of some introns have been rock stable over at least a billion years of evolutionary time. Molecular biologists have found a few genes in which the position of introns is virtually identical in all forms of life from the simplest of yeasts to human beings. They are even the same in all plants. Because the microbes and their genes first arose perhaps a billion years ago, the fact that ours are the same means that evolution must have preserved the intron positions for all this time. Obviously introns must not be mere "junk DNA." The rare but unavoidable mistakes in DNA replication during cell division should have shifted their positions. In fact, that must have happened many times, but apparently, the inheritors of those mutations did not survive. Clearly there is a reason for the gibberish sequences to stay where they are.

Taken together, all this evidence has led to one of the most important new insights in decades to the mechanisms of evolution. It is a phenomenon that Harvard molecular biologist Walter Gilbert calls *exon shuffling*. The idea grows out of the well-known fact that when a cell divides and its DNA strands are duplicated, on rare occasions they can become broken into fragments and the pieces can be reassembled in a different order. Some pieces can also be copied an extra time. Such phenomena can cause birth defects, but there is sometimes a more profound effect. Consider that the breakage can happen at any point in the DNA strand, but that if the gibberish introns are relatively long, the odds are good that the break will occur in them, where they would not break up any useful module, or exon. When the DNA pieces are spliced together again, as happens routinely in dividing cells, a gene may find it has acquired a complete new module capable of performing a recognized function.

Most often, of course, the protein that results will be useless or even harmful. But once in a while, exon shuffling will create a protein with a new ability that alters a cell's powers in some beneficial way.

The discovery of exon shuffling has yielded a powerful new argument against those who say the odds are too long that chance mutations will produce useful new proteins that give rise to evolutionary change. Chance, as Jacques Monod would have predicted, is still the operative force, but the existence of introns tips the odds heavily toward the likelihood of getting a structurally meaningful and perhaps even useful result from every throw of the genetic dice.

Whatever the evolutionary past of any protein, the next steps in its processing within the cell depend on its future. If the protein is destined for use within the cytoplasm, the ribosome that makes the protein does its work while drifting free. But if the protein is to be installed in the plasma membrane to serve as a receptor or if it is to be secreted—dumped into the space outside the cell—the ribosome is not free. It becomes attached to the surface of the vast, floppy, folded layers of the hot air balloon, the *endoplasmic reticulum* or, as cell biologists often say, the ER, pronounced one letter at a time. The ER is another of the many chambers within a cell that holds its own special mix of chemicals to carry out its part of life's metabolism. As the ribosome makes its protein, it injects it into the space inside the ER (Figure 5.11).

Unlike a hot-air balloon, there is no opening in the ER. It is completely enclosed by the same kind of phospholipid membrane that wraps most organelles. The newly forming protein gets into the ER because it carries the right address label. In other words, as the new protein is being formed by the ribosome, its first few amino acids have a sequence that is recognized by a receptor on the outer surface of the ER. The receptor pulls the head end of the still-forming amino acid chain through the ER's membrane. The code for the address label is the first exon to be translated for many proteins.

Inside the ER, the first of several chambers in which the protein will be processed, is a special chemical environment very different from that of the cytoplasm. This is where the chaperones do their work, helping to fold the protein into the desired shape. Within the ER the growing amino acid chain undergoes considerable modification. Various nonprotein molecules are attached to the raw protein, including long chains of fatty acids. Different proteins have different sites on which to hang these molecules, many becoming as decorated as a Victorian Christmas tree. Some proteins are whittled into smaller structures. Some are combined with other proteins to make a bigger complex.

Hemoglobin's four subunits, for example, are fitted together in the ER—the two alpha chains and the two beta chains interlocked into one structure. Here also each subunit of hemoglobin is given a nonprotein molecule called *heme*. Heme is an organic molecule that holds one iron atom at its center. It is the iron that takes up oxygen in the lungs and releases it elsewhere in the body. All four chains, each with its iron-holding heme, fit together to make one hemoglobin (Figure 5.12). This is

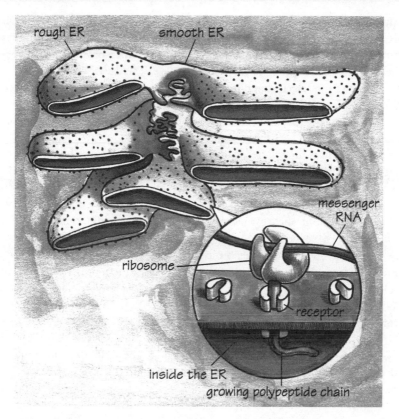

rough ER  smooth ER

messenger RNA

ribosome

receptor

inside the ER

growing polypeptide chain

**Figure 5.11** Many newly made proteins must go through a series of processing steps, beginning in the endoplasmic reticulum (ER). The ER is the vast, billowy organelle (referred to in the "living-room cell" as a folded hot-air balloon). Here a portion of the ER is shown, cut away. Each dot on the surface is a receptor where a ribosome is doing its job, reading the messenger RNA and making a chain of amino acids, which are injected into the space inside the ER.

the iron, incidentally, that is so important in a person's diet. Everyone loses a small amount every day in urine, feces, sweat, cells sloughed off from the skin, and occasionally in bleeding accidents. Women can lose larger amounts in menstrual blood. To avoid anemia, this lost iron must be replaced by dietary iron.

Within the ER lies one of the most remarkable phenomena in life. Like any other manufacturing process, protein synthesis can get fouled up. Just as workers on an automobile assembly line can miss a step and forget to install their bolt, the ribosomal assembly line that turns out proteins can miss one of its steps. Or the chain of amino acids can fold into the wrong shape. And, like any good assembly line, the cell's protein factory has a quality control program. Cell biologists have only just discovered that it exists and are just beginning to study it, but it is already clear that there is an apparatus within the cell that can detect misshapen or otherwise defective proteins during the manufacturing process, pull them out of production, "melt" the rejects down for their raw materials, and send those back for recycling.

A cell that lacked the ability to catch mistakes could end up, for example, installing useless receptors on its surface. They would fail to take in the substances the cell needs to survive, or perhaps the cell might never recognize a hormone that is supposed to trigger the cell to do its job for the body. Obviously, the life of the cell and the health of the person depend on cells having properly functioning proteins. It is even conceivable that without such a quality control program, life could not have evolved much beyond the one-celled stage. A single cell that turned out misshapen proteins might die, but that would not harm the species. But a multicelled organism depends on the successful functioning of trillions of cells simultaneously, and if some of those cells must live for many decades, as is the case with human muscle and nerve cells, there is much less room for error.

The first stage of the quality control process happens within the ER. One of the best-studied examples involves a very complex receptor protein that is essential to the immune system. It is manufactured by the white blood cells called T cells and does its job as a receptor in the T cell's outer membrane. The receptor is a modular protein made of seven different

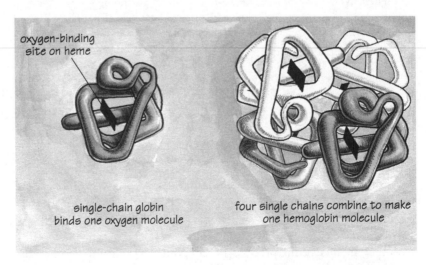

oxygen-binding
site on heme

single-chain globin
binds one oxygen molecule

four single chains combine to make
one hemoglobin molecule

**Figure 5.12** Inside the ER is where hemoglobin's four subunits fold into their proper shape and then come together to make a hemoglobin complex. At left is a globin formed from a single chain that has folded and enclosed a heme group, an iron-based molecule where oxygen binds and from which it is released.

smaller proteins. Each of those seven is the product not just of different exons but of different genes, and in some cases the genes are even on different chromosomes. So to make the receptor, seven ribosomes must manufacture seven amino acid chains, injecting each one independently into the ER. Inside the ER the chains must find one another and lock together into the right configuration. The result, as in other receptors, has three regions: one that sticks out of the cell to meet the molecule to be received; a hydrophobic region that passes through the cell membrane, keeping the receptor properly anchored in it; and a region that sticks into the cell to carry out the appropriate response when the outer portion receives the right molecule.

Experiments with cultured T cells show that it takes about thirty minutes for batches of the seven newly made modules to be assembled into complete receptor complexes. Normally these complexes are then shipped from the ER to the *Golgi apparatus*—yet another processing facility that, in the living-room cell, is represented by the drifting stacks of deflated beanbag chairs. From there the receptors are dispatched outward for installation in the plasma membrane (Figure 5.13).

The experiments also led to a surprising finding—that partially made receptors containing fewer than all seven components almost never are installed in the plasma membrane. In fact, they don't usually even reach the Golgi apparatus, which most cell biologists call simply the Golgi. (It is usually capitalized because it is named for Camillo Golgi, the Italian physician who described the organelle in the late nineteenth century.) The mismade proteins remain trapped in the ER. Biologists know the receptors are missing a module because the experiments have been done with cultured cells from which one of the seven genes was removed. If any of the seven component proteins is missing, the others will assemble as far as they can go but the resulting incomplete complex is blocked from exiting the ER.

So, who does the inspecting? How does the ER decide whether a protein is properly made? Thousands of different kinds of proteins are processed through the ER; can there be a specific quality control mechanism for each?

One clue has come from the discovery of a protein that resides permanently inside the ER and which binds to certain newly made amino acid chains before they can fold. It is called, prosaically, binding protein, or BiP, which is, of course, pronounced bip. Researchers have found that in some cases BiP binds to an unfolded protein while it is being injected into the ER but then breaks away as the protein achieves its fully folded form. It looks as if BiP is not terribly choosy about the proteins it binds to, but always binding to any region of a protein that is hydrophobic, or water-repelling. Many proteins contain such regions, but when the amino acid chain is completely folded, the hydrophobic region is supposed to lie

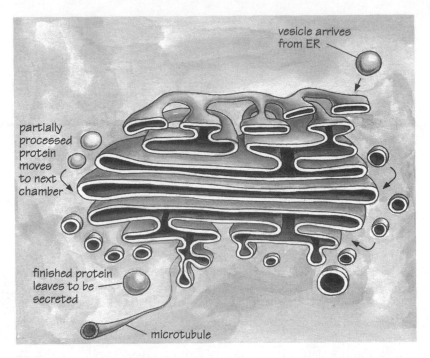

**Figure 5.13**    After the endoplasmic reticulum, many partially processed proteins move to the Golgi apparatus for further work. Each Golgi is a stack of separate compartments. At upper right is a vesicle carrying some of these partially completed proteins from the ER. It will fuse with the membrane of the Golgi's first compartment dumping its cargo inside. When the first part of the work is complete, vesicles will bud off the first compartment and carry the protein to the next. When completed, the vesicles from the last compartment are transported away, probably hauled by motor molecules along microtubules.

buried deep inside the overall configuration, in a position where other molecules can't "see" it. The outer faces of most proteins are, by contrast, hydrophilic, or water-loving.

So how would BiP keep an unfolded or improperly folded protein in the ER? Possibly because its own amino acid sequence has a tail (the head binds to the newly made proteins) that sticks to some receptor facing the inside of the ER. Some researchers have speculated that this is the case because of a curious similarity among BiP and several other proteins known to be permanent residents of the ER. They all have tails composed of the following sequence of four amino acids: lysine–aspartic acid–glutamic acid–leucine. This similarity suggests a similar function for the sequence, and since all the proteins that have this sequence stay in the ER, it is simply a logical guess that the function could be to anchor BiP somehow to the inside face of the ER.

If this speculation is right, it would turn out once again that a process within the cell that seems at first to be so technically daunting can actually be carried out by a surprisingly simple mechanism. The quality control inspector would be a molecule that keeps one end attached to the inside of the ER while its other end waves about trying to grab onto any surface with a general chemical property (hydrophobicity, in this case) that ought not to be on the outside of a properly made protein.

But the story is surely not this simple. Many incomplete protein complexes that remain trapped in the ER do not appear to interact with any known component of the ER. The T-cell receptor, for example, does not appear to bind to BiP or any other ER denizen.

When the ER has finished its modifications of the new protein, it ships the result to the Golgi. A tiny patch of ER membrane bulges out like a soap bubble from a child's toy and breaks loose. The little bubble, or vesicle, contains ER-modified proteins and moves to the cell's next processing station, the Golgi. Actually, the Golgi consists of a series of chambers, each a flattened sac stacked on top of the previous sac. The movement of the vesicle is not passive. Some kind of motor molecule transports it.

Although the shipping of vesicles is well known, it was only during the late 1980s that scientists began to learn the mechanism that made it happen. As any one who has played with soap bubbles knows, bubbles don't just suddenly sprout little buds. Something has to make that happen, some mechanical force. The mechanism that does it in the ER turns out to be very much like the one described in the previous chapter where the outer membrane of a cell would form a clathrin-coated pit to take in some molecule from outside the cell. The binding of the star-shaped clathrins, sticking to each other and to the membrane, form a cage that pulls the membrane into a pit that finally closes into a vesicle completely enclosed by interlocked clathrins.

Jim Rothman and his colleagues at Sloan-Kettering Memorial Cancer Research Center have recently found that something similar happens in the ER but that the agent is not clathrin. It is a different set of proteins, but in another example of how evolution has hit upon similar mechanisms to carry out similar jobs within the cell, the proteins seem to work much like clathrin. They clamp themselves on to the ER, perhaps held there by some tail of protein sticking out of the ER membrane, and they clamp to each other. Because of their shapes and the angles formed as they lock into one another, they pull the ER membrane into a bulge and, finally, pinch it off into a spherical vesicle. Rothman has found that the vesicles are always the exactly same size, another indication that their form is determined by the way the coating proteins lock together. In the living-room cell the vesicles would be a little larger than a marble. All around the edges of the folds of the hot air balloon that constitute the ER, marble-sized balls would bulge out, pinch off, and head for one or another of the stacks of deflated beanbag chairs, the living-room Golgi. They go, transported by motor molecules, to the first bag in the stack.

Upon reaching the Golgi, a protein sticking out of the vesicle's surface interacts with others protruding from the Golgi. If they fit together, the vesicle has found the right destination and, after some further interaction between molecules on the surfaces, the membranes of vesicle and Golgi fuse, automatically dumping the cargo of partially processed proteins into the Golgi. Somewhere in this sequence of events, the protein cage fell away.

Rothman notes that the fusion of membranes doesn't just happen. "You can push two membrane-bound vesicles together as long as you want and they won't fuse," he says. "There has to be some other agent that comes into play and makes fusion happen."

Rothman points out that cells are jammed with vesicles and organelles, all bounded by essentially the same kind of fluid-mosaic membranes and that fusions are happening all the time all over the cell. "But just stop and think about this. Imagine all the fusion molecules that have to be running around in the cell. But only certain membranes can be allowed to fuse. Imagine the damage if a lysosome full of digestive enzymes fused with a nucleus. The governance of membrane fusion must be one of the most tightly regulated phenomena in the entire cell." Without that strict regulation, one of the fundamentals of life—the compartmentalization of chemical reactions—would be impossible.

The Golgi apparatus is a kind of assembly line that receives proteins from the ER, which may already have fastened fats to them at various points, and processes them further, one step at a time. The Golgi typically consists of four to six chambers. The protein gets modified in one chamber—some of the carbohydrate chains, for example, may be trimmed or remodeled—and then sent to the next chamber for further processing.

The goal is to modify each protein as necessary to suit its function and then sort the proteins by type and by destination, either in the cell membrane or outside the cell.

There is evidence that the Golgi, like the ER, has its own quality control program. If, for example, the seven-moduled T-cell receptor is made so that it lacks either of a certain two of the modules, the complexes do get shipped from the ER to the Golgi, but there they are blocked. A faulty protein complex may slip past the ER's inspectors, but the Golgi has a backup system for catching mistakes.

By the time a protein leaves the Golgi's last sac, it is fully formed and ready to go to work. A vesicle buds off the edge of the last Golgi sac and heads for some destination. If the protein is to be secreted, its vesicle is transported to the plasma membrane and fuses with it, dumping the contents outside the cell (Figure 5.14). If, for example, the product is insulin, the molecule soon diffuses into the bloodstream to begin its work in helping the body metabolize blood sugar. If, on the other hand, the product is a receptor intended for installation in the cell's surface, it takes up its position in the membrane of the Golgi, and when the last Golgi vesicle fuses with the plasma membrane, the receptor is automatically in position to begin its job as a gatekeeper for the cell.

Tragically, in the case of sickle-cell hemoglobin, no quality control mechanism detects the flaw of a single errant amino acid slipped in among the 861 that dictate the structure of the alpha and beta chains. The mutation leaves the molecule looking perfectly fine to all the molecular inspectors, and the cell dutifully relies on it to perform the critical job of carrying oxygen to all the body's cells.

Although sickle-cell anemia is probably the most intimately known of all genetic diseases, the knowledge has not given rise to many advances in treatment of those afflicted with it. This fact underscores a point often overlooked in all the hoopla about finding the gene for this disease or the gene for that condition. Knowing the gene tells you little or nothing about what the protein does. Additional research, based as much on the study of living cells as on the analysis of genetic codes, is needed to reveal what role the wrong molecule or the right molecule is playing in the process of life. Even that, however, is not always enough. As we have seen, the function and behavior of the faulty hemoglobin molecule are well known. Other

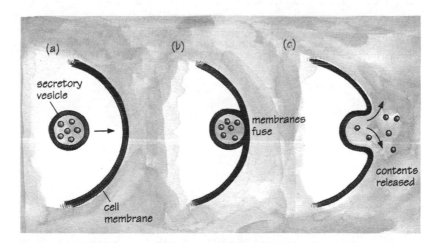

**Figure 5.14** (a) Proteins that are to be secreted arrive at the cell's outer membrane in a vesicle. (b) Receptors on the vesicle's outer surface interact with others on the cell membrane, causing the two membranes to fuse. (c) This automatically releases the vesicle's cargo.

lines of research, however, have led to treatments that make the lives of many victims easier.

One relatively new treatment does make use of a key finding of cell and molecular biology. As it happens all people are endowed with two different forms of hemoglobin, one that the body makes and uses chiefly during embryonic and fetal development and another that is manufactured and used after birth. There are different sets of genes for fetal hemoglobin and adult hemoglobin. Fetal hemoglobin is believed to be more vigorous at taking up oxygen, which is a more difficult job for the fetus than it is for a born person. The fetus must get its oxygen in the placenta, grabbing it away from the mother's adult hemoglobin. After birth, the gene-regulating molecules change, and the gene for fetal hemoglobin is gradually slowed down over a period of months and finally turned off as the gene for adult hemoglobin is phased in.

As it happens, the fetal hemoglobin gene is perfectly good even in people with sickle-cell anemia. If that gene could somehow be switched back on, researchers thought, it should make hemoglobin that can't polymerize, or lose its oxygen-carrying ability, or warp red blood cells into poorly flowing sickles. A way has been found to do this, using a form of urea, the nitrogen-rich molecule formed in the liver as the end product of protein breakdown and which is excreted in urine. If sickle-cell patients are given a drug made from urea, their fetal hemoglobin genes are switched back on. They make enough fetal hemoglobin that sickling is largely blocked, even though the faulty adult hemoglobin gene is still working. Urea's exact mechanism of action has not been worked out fully but it clearly sets in motion some kind of process that reawakens the long-dormant gene.

Research on sickle-cell anemia has also brought a wealth of other knowledge of how intimately the tiniest details of the genetic code carried in those long threads of DNA can affect the lives not just of cells but of whole human beings.

## CHAPTER 6

Item: You accidentally bite your lip or cut yourself shaving. Or, let's say, a child skins a knee. In a matter of days, the wound heals and the tissue is as good as ever. Even major injuries, given enough time, commonly heal.

Item: A cancer arises somewhere in the body. It starts indetectably small, a single cell gone awry, but the cell multiplies. When the mass of proliferating cells is big enough, surgeons may remove the tumor and hope to cure the patient, but all too often, it already will have seeded new tumors throughout the body, hopelessly beyond the surgeon's knife.

Item: From a single cell, a fertilized human egg, an embryo develops. Within months it becomes a baby weighing several pounds and made up of several trillion cells.

Underlying all these commonplace events is one of the most extraordinary natural phenomena known to science—the ability of one living cell to transform itself into two living cells.

When the body is injured, certain special cells lurking among those in the damaged tissue spring into action and begin dividing (or multiplying, if you prefer), spawning new cells that take the places of the destroyed ones. When cancer begins, it is but a single normal cell that has lost its stability. Somehow, the mechanism that times its cycle of normal cell divisions goes haywire. Heedless of a built-in "stop" signal that keeps normal cells in check, the renegade cancer cell multiplies into a huge mass and may slough off cells that take their lawless ways to other parts of the body. And, again, cell division is essential to transforming a microscopic speck in a mother's womb into a fully developed human being.

Autonomous motion, as we have seen, is one of life's hallmarks, but it is cell division that most dramatically certifies life as a phenomenon utterly unlike any other. Lifeless machines like steam engines and automobiles can be made to huff and puff and move about, but they cannot be made to spawn new machines in their own image. Alone among all the known objects in the universe, it is the living cell that can create a duplicate of itself. No outside force dictates the process. The power and the plans lie within each cell, implicit in its molecules. At various stages in the growth of many organisms, certain of their cells give up the power to keep dividing, but until that happens, the relentless wheel of cell division keeps rolling. As François Jacob, a French molecular biologist who shared a Nobel prize for discoveries about gene regulation, put it, the "dream" of every cell is to become two cells. This is the sine qua non of life, the sine qua non of the living cell. So fundamental is this cellular dream, in fact, that cells which had been obedient members of the vast republic of the human body sometimes rebel and sacrifice their host society to berserk ambition. This is cancer, and as we shall see in a later chapter, it is the result of releasing the brakes that normally halt cell division in most of the body's tissues, or that keep its timing under strict cadence.

Cell division has also played a fundamental role in evolution. It unites all forms of life on Earth—the apple to the aardvark, the bacterium to the bison, the carp to the crow. At the dawn of life, when the first primordial cell threw itself on the mercy of a hostile new planet, it was cell division that ensured the future. That lone cell would surely perish, but if it could have progeny, flung far enough, its kind would survive. More than that, it is cell division that provides the opportunity for most new evolutionary changes to be created and installed in the genes of a new generation, a new, altered form of life that may grow up with a new form or function.

And it is cell division that has kept life's fire burning for perhaps four billion years, relaying its spark from one parent cell to two daughter cells, from the archaic microbes to their multicelled descendants the worms, from the primitive fishes to the amphibious pioneers of the land, from the wide-eyed little primates flitting about the trees to the first creatures capable of looking back and wondering how they came to be.

"Cell division is truly one of the great mysteries of life," says Shinya Inoue, a pioneer in the modern study of cell division at the Marine Biological Laboratory. Inoue, who developed some of today's methods of electronically enhancing microscope images of living cells, has specialized in the study of cell division. "If we could understand it all, we would know life at a very profound level indeed."

For one thing, we would know an event that happens at least 10 million times in an adult human body every second. Except for muscle and nerve, all of the body's cell types are in a state of continual death and renewal. Cells grow old and die. Cells die when they fail to receive correct signals from their neighbors. Or cells are damaged in the course of their duties and die. In a normally healthy body, none of this is a problem. As in the wounded skin, there are special damage-repair cells and progenitor cells that wait for this moment. When an old cell goes, these reserve members of the community divide, producing two cells—one that quickly develops the specialized characteristics to take the place of the lost cell and one to remain behind as an unspecialized cell, ready if needed again.

The fastest to turn over are cells lining the intestinal tract, which live for only a day or so in the harsh environment of digestive juices and the grinding-by of indigestible matter before dying. Skin cells cycle in weeks—old, dead ones sloughing off at the surface as new ones are born in a layer deeper inside. (The gut lining and the skin are, incidentally, the tissues most damaged by cancer drugs because those chemicals interfere with cell division, and the more a tissue's cells divide, the more it is hurt by the drugs.) Muscle cells normally do not divide, but injuries can activate special dormant cells to multiply and rebuild damaged muscles. Neurons, the cells of the brain as well as the sensors and wires of the nervous system, are thought never to divide once the full complement is achieved in childhood. New evidence, however, is indicating that some, after all, can divide in special cases.

These differences in frequency of cell division are directly linked to the propensity of a given cell type to become cancerous. The more often a cell normally divides, the greater its chances of spawning a cancerous daughter cell. Thus, cancers of the digestive tract and the skin are the most common. Cancers of muscle are far less common and nerve cancer is almost unheard of. (Brain tumors almost always involve not the neurons but other non-nerve cells found in the brain, the so-called glial cells, whose job is to feed and care for neurons.)

Even within a tissue of frequently dividing cells such as the outer layer of skin—called the epidermis—not all the cells divide. The tissue contains a population of what are essentially embryonic cells whose division produces a new embryonic cell and another cell that develops into the mature form. Skin's embryonic cells form a one-cell-thick layer deep below the surface. There these so-called basal cells divide repeatedly, producing one cell that remains behind as a basal cell to divide again and another destined to move toward the surface to form a tough, protective coating for the body.

In every tissue, there are regulatory mechanisms that tell a cell when to divide and when not to divide. Failures in this mechanism lead to some of the most catastrophic diseases. Comparable errors during embryonic growth can cause a wide variety of birth defects. Brain cells, for example, are supposed to multiply until early childhood and then stop. But, if the "stop" signal arrives before birth, the brain fails to develop fully and the child may be born with little or no brain. Unregulated cell division, of course, can become cancer. Experiments have shown that when normal human cells are dropped into a container of nutrient broth, they come to rest on the bottom and then begin to crawl about and multiply until they touch one another, covering the bottom of the container with a layer of cells. Contact with surrounding cells somehow sends a message to the nucleus. The cells stop dividing, a phenomenon known as contact inhibition. However, when cells are taken from a tumor and put in culture, there is no contact inhibition. They multiply until they form a layer covering the bottom of the container, and then they keep on multiplying. The proliferating cells just pile themselves into higher and higher heaps. Whatever it is that tells a cell to stop dividing can no longer get its message to a cancer cell. In a little plastic dish, or a cell-culture flask, the relentlessly dividing cells try to create a tumor.

Since the earliest days of the microscope, scientists have watched cells multiply and tried to fathom what was happening (Figure 6.1). In the early nineteenth century, when cell theory was just beginning to be established, ideas varied enormously. Some researchers saw analogies to the ways in which whole animals reproduce. Others, for example, held that new cells began as embryonic cells within an existing cell and then somehow emerged much as a baby is born. They could even point to structures within cells (we now know they are organelles) and say they were embryonic cells. Still others insisted that the process was more like the way plants grew. They saw a bud forming on one side of a cell and judged it to be an infant cell that would grow to adult proportions.

By the mid-nineteenth century several researchers had learned to fertilize frog eggs under the microscope. They watched the egg change, cleaving into two cells, and then two into four cells.

Missing from these early views was any perception of what is known today to be the most fundamental part of cell division—the faithful copying of the parent cell's entire set of chromosomes, so that the two daughter cells could inherit identical genetic endowments. In those days everyone understood quite well that like came from like, that cows gave birth to cows and humans to humans. Breeders of show plants and farm animals knew that special traits of the parents could be passed on to the offspring with considerable reliability. But they had no idea how this happened.

Today the phenomenon is understood in astonishing intimacy. To be sure, pieces of the puzzle are missing, but so much of the picture is in place that it is clear that when the missing information is found, it is not likely to change the overall scene. Moreover, research on cell division is revealing the causes not only of cancer but of some forms of mental retardation, birth defects, and a host of other diseases.

At the heart of cell division is mitosis, the process of duplicating the chromosomes and distributing a complete set to each of the two daughter cells. Organelles that exist in many copies, such as lysosomes or mitochondria, are simply apportioned to the daughter cells as the original cell pinches in two. As the daughter cells grow in size, they will make more organelles.

The first job of a dividing cell is to make a second set of chromosomes so that each of the two daughter cells can have one complete set. Because

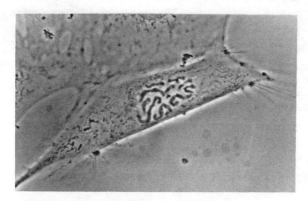

Figure 6.1a

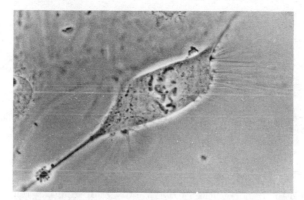

Figure 6.1b

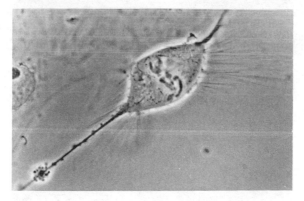

Figure 6.1c

**Figure 6.1**  A cell growing in a laboratory culture dish divides. In the first view (a), condensed chromosomes can be seen in the nucleus. Then as the cell begins to round up (b and c), the chromosomes assemble on the so-called metaphase plate. The chromosomes pull apart into two identical groups (d), and the cell begins to pinch in two (e, f, g). Finally, the new daughter cells spread out again (h) to resume life as functioning cells. Using a scanning electron microscope, which can see only the outer surfaces, a dividing cell looks like this (i), a stage that corresponds to (e).   (*Figure 6.1 continued on page 122*)

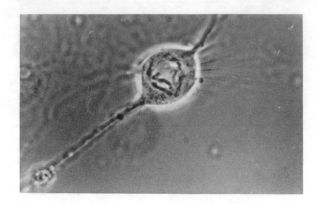

Figure 6.1d

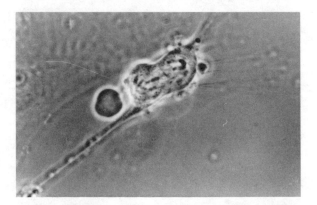

Figure 6.1e

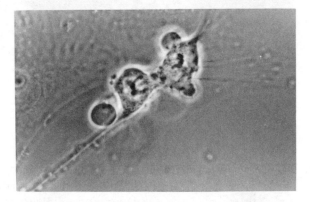

Figure 6.1f

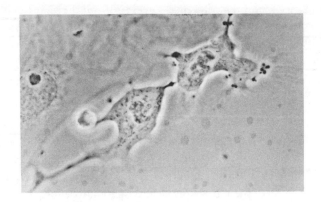

Figure 6.1g

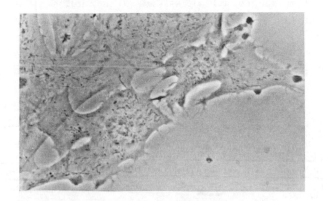

Figure 6.1h

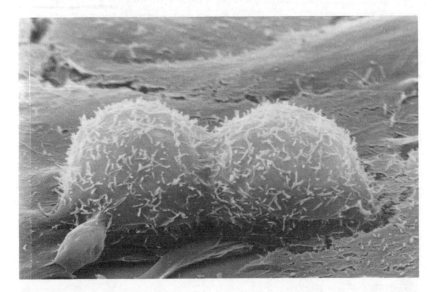

Figure 6.1i

DNA replication and the subsequent events of cell division force all of a cell's genes to be dormant for a period of hours, all the genetic instructions must be issued in advance. The requisite genes command the synthesis of a suite of enzymes and mechanical devices that will not only make a duplicate set of chromosomes but will sort them into two identical sets and then cause the parent cell to split precisely between the two chromosome sets.

During mitosis, the chromosomes must give up their usual filamentous form (the miles of thread festooning the inside of the VW nucleus) and shrivel into sausage-shaped blobs (fat Xs just a few inches long at the VW scale). The genes are like science-fiction space travelers who, having built a spacecraft to carry them in suspended animation to some distant galaxy, abandon their normal, active form and enter an inert state until cell division is complete and the genes have been delivered safely into the nuclei of two daughter cells. Only then do they unshrivel and resume their command. The entire sequence of events is regulated by the biochemical equivalent of the controller on a clothes-washing machine. Just as the washer has an apparatus that makes sure to fill the tub with water before it starts agitating and to drain that water before going to the rinse and spin stages, so a dividing cell has signals and controllers to stop and start each stage of cell division. We'll look at the mechanism in more detail later, but for now the point is that there are very specific signals that come (if everything is working right) at predetermined checkpoints.

Once a complete stockpile of enzymes and structural proteins is in place, but before the chromosomes condense into their inert form, the chromosomes must be replicated—so there are two identical sets. When the signal comes, a suite of DNA-processing enzymes, the main one being DNA polymerase, go into action. At a series of points along each of the human cell's forty-six chromosomes, special enzymes split apart the two strands of the double helix. Each original strand will then be used as a template along which to assemble a new second strand, restoring DNA's normal double-stranded form. The result will be two identical double strands, each composed of one strand from the original double strand and a newly made complementary strand bound to the original along their entire lengths (Figure 6.2).

This is possible because the two strands are complementary—each is, in a sense, the mirror image, or a mold, of the other. As we saw when DNA was expressing its message in Chapter 5, all strands of DNA are made of four kinds of subunits, called nucleotides and abbreviated A, T, G, and C. The chemistry of nucleotides allows a single strand to contain any sequence, but the opposite strand must always be of a complementary sequence. The opposite strand's complementarity is like that of messenger RNA when it transcribes the genetic code. In other words, if there is an A at a given position in one strand, only a T can go into position opposite it on the complementary strand and vice versa. In the same way, C and G are a complementary pair. For example, if one strand of a double helix has the sequence ACGT, the complementary sequence on the opposite strand will be TGCA.

DNA replication is carried out in the presence of loose nucleotides. While some enzymes are "unzipping" the double strand, others are bringing loose nucleotides into position to assemble the complementary strand opposite each of the two original strands.

Accuracy in this process is essential. If just one nucleotide were omitted, for example, or if an extra one were inserted, it would throw off the

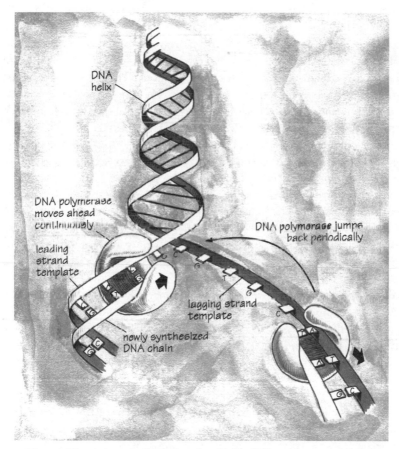

**Figure 6.2** A strand of DNA—the double helix at the top—is being duplicated. The blob on the left, representing a molecule of DNA polymerase, unzips the two original strands, exposing complementary rows of bases, labeled A, T, G or C. It also guides synthesis of a new, complementary strand. On the other original strand, another polymerase molecule works away from the fork in the original strand and then jumps back. The rather arcane reason is explained in the text.

rest of the sequence because it would shift the *reading frame.* This is the group of three nucleotides that DNA polymerase is reading to specify one amino acid in the resulting protein. Nucleotides run one after the other with no spacing between. So when the cell is to make a protein and comes to read the gene, it counts off the letters of the code in threes from the start *Codons* of the gene. An omitted or inserted nucleotide shifts the reading frame, altering the meaning of every single triplet codon. It would be like shifting the space between words in an English sentence one step away from where it should be. For example: Itw ouldr eadl iket his.

Though the cell's DNA copying system is highly accurate, it does make mistakes. If it made too many, however, a cell's genetic program would become so scrambled that it would not work and the cell would die. To prevent this, nature has evolved a molecule to act as proofreader to catch mistakes. It is DNA polymerase, the same versatile molecule that assembles the complementary strand. Every time a loose nucleotide tries to fit itself onto the growing new strand, a portion of the DNA polymerase molecule checks it for a proper fit with its supposed complement on the opposite strand. If the fit is wrong, say a G trying to bind to a T or an A, the enzyme automatically yanks the errant nucleotide.

DNA synthesis must be highly accurate. Every error that slips through is a mutation that could doom the cell, and if the error happens to be in a certain kind of gene, it may turn it into a cancer gene. Cancer researchers have recently discovered that a key gene for colon cancer codes for an enzyme whose job is to repair defects in DNA. Other mutations in DNA may be repaired before they do harm, but if the error is in the repair enzyme itself, prospects are dim. This enzyme travels along the newly made double helix within minutes of DNA synthesis and looks for errors—places where the opposing nucleotides are not complementary. Then it essentially takes out the bad nucleotide and puts in one that correctly complements the original strand. How does it know which strand is the original one? Nobody knows yet, but it is known that bacteria have a similar DNA repair enzyme that checks the DNA for attached clusters of atoms called methyl groups, which are only on the old strand. (Methyl groups will be added to the new strand but not for about ten minutes, giving the repair enzyme only a brief window in which to figure out which strand is which.) Higher organisms do not have methyls attached to their original DNA strands but may have some similar marker.

The repair enzyme does not replace only the mismatched nucleotide. It rips out a long segment of the strand containing the error (from one methyl group to the next in bacteria), and DNA polymerase synthesizes the replacement (Figure 6.3).

Molecular biologists have calculated that if DNA replication were not as accurate as it is, evolution probably could not have produced anything more complex than an insect. That's because errors are made in DNA replication at a high enough rate that, without some method of proofreading and error correction, essential genes in a more complex organism would be hopelessly mangled. Here's how they figure it: DNA replication makes a mistake in about 1 out of every 10,000 nucleotides it adds to build the new strand. The proofreading and error-correcting ability of DNA polymerase itself, cuts the actual error rate to 1 in a million to 1 in 10 million. That's good enough for a simple organism that has relatively few genes. So few cells will experience a single copying error that the organism should live just fine. DNA polymerase alone could get them all copied, and they stand a good chance of escaping error. But for a human body, which requires the proper functioning of many more genes in many more cells, accuracy must be higher still. That's why the repair genes are essential. They cut the final error rate to 1 in a billion nucleotides. And, because most of those errors will be in the larger stretches of noncritical DNA, most cells will inherit fully functional sets of genes.

A typical human cell, containing 50,000 to 100,000 genes made of 3 billion nucleotide pairs, takes about seven hours to make one copy of all its genes. This is a fairly long time, but after all, it is the equivalent of reading a thousand 500-page books in which each printed letter represents one nucleotide in a cell. At the end of the process, the original complement of

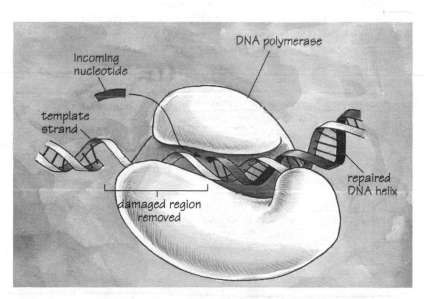

**Figure 6.3** A molecule of the enzyme DNA polymerase enbraces a damaged region of DNA, snips out the bad parts, and replaces them with good nucleotides. The enzyme uses the remaining complementary strand to determine the proper nucleotides (A, T, G, or C) to go into position on the opposite strand.

forty-six chromosomes has become ninety-two—that is, two identical sets of forty-six each.

For all the elegance of the DNA double helix and the precision of replication, the process must contend with some awkward architectural problems. To solve them, evolution has produced mechanisms that are both clever and, in a sense, comical, if molecules can be amusing.

The first problem is the double helix itself. The famous shape is necessary given the way its components link together, but the shape actually turns out to be a big problem for the cell. To copy the DNA sequence, the two strands must be separated, "unzipped." But the strands are wrapped around one another, wrapping thousands of times around each other along a very great length. Imagine trying to unwind the strands of an ordinary rope, which is usually a triple helix of three smaller ropes wrapped around one another. You can flop one strand over another for a few feet, but if you have miles of rope to unwind, you will soon have three virtually impossible lengths of frayed ends to manage. The dividing cell has exactly the same problem. So how does it cope?

The cell does what the human would wish to do with the rope. It simply chops the long rope into shorter pieces, moves one loose end around the other, and splices the cut ends back together again. Yet another set of special enzymes, of course, does the job. Unlike spliced ropes, which have a thick bulge at the splice, spliced DNA is as good as new. The cut ends simply stick together again exactly as they were.

The double helix poses a second architectural problem, one to which evolution would appear to have chosen a very awkward solution, or at least a solution far more cumbersome than one that a human might try to design.

Each single strand of DNA has a polarity, or directionality, to its so-called backbone, the chain of alternating sugars and phosphates. This means, for one thing, that a gene can be read in only one direction, just as English is read in only one direction. This also means that DNA polymerase, the enzyme that assembles the new strand of DNA (complementary to the old strand), can work in only one direction. The problem arises from the fact that the complementary strands of the double helix have opposite polarity. They are like trains going in opposite directions on parallel tracks. In other words, one strand can be read (and replicated by DNA polymerase) only from left to right while the complementary strand can be read (and replicated by DNA polymerase) only in the opposite direction, from right to left.

If the two strands were completely unzipped, there would be no problem. One molecule of DNA polymerase would do its job by moving from left to right on one strand while another started at the opposite end of the other strand and moved from right to left. But in living cells the two strands are bound to each other until the DNA polymerase molecule comes along and unzips them.

A human being trying to design a solution would probably suggest simply making a second form of DNA polymerase that could read DNA in the opposite direction. After all, there's nothing magic about the directionality. In principle, some alteration of the DNA polymerase should do the trick. Nature's solution to the problem, however, is worthy of an "I Love Lucy" television comedy.

The double helix is unzipped, a little at a time, let's say in a northward direction, and opposite lengths of each strand are exposed for copying. One DNA polymerase molecule moves smoothly north along one strand—the same direction as the unzipping. On the opposite strand, however, another DNA polymerase starts its work at a point on the exposed sequence but can move only toward the south—while the northbound unzipping continues, exposing more of the southbound strand. After the southbound DNA polymerase has assembled a short segment of new complementary strand, about a thousand bases long, it jumps off the DNA and moves back north toward the unzipping point, which by now has exposed about a thousand more bases on the southbound strand. Then the DNA polymerase starts working again, making a new segment of DNA that will be linked to the previously assembled segment.

So, while one strand of the old helix is processed smoothly and continuously, the other is subject to frenetic, back-and-forth activity that makes segments of new DNA and splices them together. It seems awkward, but, of course, the results are just fine. And that is all evolution cares about.

But the architectural problems of DNA replication don't stop there. A new problem arises when the process reaches the end of the double helix, the very tip of the chromosome. The strand being processed smoothly has no problem. But not the other strand. Every time the DNA polymerase must double back to a new starting point on its strand, it needs to find a little DNA on each side of itself before it can go to work. This means that when the back-and-forth strand is being processed and DNA polymerase is making its last jump toward the end of the strand, it cannot process the very last few bases. The complementary strand it is assembling will be a little shorter than the original.

It is easy to imagine that with each round of cell division, the tips of the chromosomes would be lost and the whole chromosome would become shorter and shorter. Sooner or later, essential genes would be chipped away and the cells would die. The fact that cells have been dividing happily for billions of years obviously means that they must have some mechanism for avoiding this fate.

Molecular biologists have long had a clue in the fact that the ends of chromosomes don't contain genes in the usual sense but instead have scores of short repeating sequences. In human cells, for example, the sequence is TTAGGG—the same six-base sequence again and again. This obviously means that the ends of chromosomes are not genes but have some other function. But the mystery of the chromosome's survival remained because these sequences should have been lost just as easily as any other.

Elizabeth Blackburn, at the University of California at San Francisco, and her colleagues have discovered the answer at least for one-celled organisms and the sex cells of higher species. It is yet another mechanism needed to compensate for having a double helix of strands with opposite polarity. It turns out that there is a special enzyme made not just of protein but also containing a length of RNA. In the RNA is the complementary sequence for the strange six-base code repeated so many times at the end of DNA strands—AAUCCC, plus part of the next repeat.

Blackburn's experiments show that when DNA polymerase comes to the end of its herky-jerky, north-south progress, it does indeed fail to process the last few nucleotides of its strand. But the only thing that is lost is part of the segment of GGGTTA repeats. Before DNA replication is finished, however, Blackburn's enzyme comes into play and adds a few new TTAGGG repeats to the end of the chromosome. It does so by using its own RNA as the template. As a result, there will always be a safety margin between the endmost true gene and the tip of the DNA strand.

Toward the end of the period in which all the enzymes are coping with the structural problems of the double helix, the finished ends of new double-stranded DNAs begin shriveling, coiling, and supercoiling into the compact, or condensed, form. Again, as seems to be the case in all of cell biology, if there is specific motion, there is some kind of motor molecule to do the work. For the job of condensing the DNA filaments into compact forms, cells have molecules that grab free loops of DNA and pull them into tighter form, something like a person coiling a length of rope.

At this stage the newly twinned chromosomes remain attached at one point along their lengths, forming the familiar X shape. In some chromosome pairs, the attachment is so close to one end that the X looks like a V or Y. To begin with, the miles of DNA thread in the VW Beetle nucleus of the living-room cell are looped around pea-sized protein balls, forming what look like beads on a string. They remain in this form almost all the time. But now the string of beads coils on itself again and yet again, pulling itself into an ever more compact shape. Finally, a fully condensed chromosome would be as thick as a person's arm or, more aptly, as thick as a medium-sized boa constrictor. Their lengths would vary, as would those of snakes. As it happens, chromosomes at this stage also move slowly over and around one another. So, in the midst of mitosis, when all ninety-two chromosomes are at their fattest, the VW nucleus would contain a tangled mass of ninety-two writhing snakes. Or, to be accurate, it would contain forty-six pairs of identical twin snakes, members of each pair somehow glued together somewhere along their lengths.

Once two identical sets of chromosomes have been created and condensed into the boa constrictors, they must be sorted, one member of each pair destined for its own nucleus. To do this, the cell constructs a cagelike machine, an assemblage of molecules that will surround the ninety-two chromosomes and reach in from opposite sides with long fibers to grab the forty-six members of each set. For generations biologists thought that this machine, called a spindle, pulled the chromosome sets to opposite sides of the mother cell. In the late 1980s, however, several researchers showed that it is probably the chromosomes themselves that do the work. They appear to pull themselves along the fibers toward opposite ends of the cell. Once all the chromosomes of each set have reached their destination a new nuclear membrane will form around each (Figure 6.4). We'll look at all these steps in more detail.

Biologists were once puzzled as to how the mitosis machine could select exactly the right forty-six chromosomes destined for each new nucleus. The apparatus has to work right every time, for if it doesn't, one cell could get double the normal number of, say, Chromosome 14's, and the other cell will get none. This does happen sometimes in the cells that are making sperm or eggs, and if they get fertilized, serious genetic diseases usually arise. The sorting process is something like two people reaching blindfolded into a drawer with forty-six different pairs of socks and each pulling out only one member of every pair. How could a mere assemblage of molecules approach a tangled hodgepodge of ninety-two objects, select only the right ones, and put them into separate piles? The process is elegantly simple.

Here's what happens: The chromosome-separating machine, or spindle, originates near two small organelles called centrioles. Under the microscope a centriole looks like a short bundle of hollow tubes. In the living-room cell, each would be the size of a soup can. The centrioles lie within fuzzy-looking balls of unknown composition called centrosomes.

It is still a mystery how centrioles and centrosomes work, but when mitosis is about to begin, the two centrioles (the original replicated itself earlier), which hover together near the nucleus, move apart and take up positions on opposite sides of the nucleus. It is a mystery how they move. Actually the centrioles began traveling while the replicated chromosomes were condensing from their threadlike form. At the same time, the membranous nuclear envelope, which had enclosed the chromosomes since their last cell division, was breaking down and its components were stored in tiny packets for reuse in building the two new nuclei. And scores of filaments, called spindle fibers, were sprouting from the two centrioles and growing toward the chromosomes, which lay between them. In the

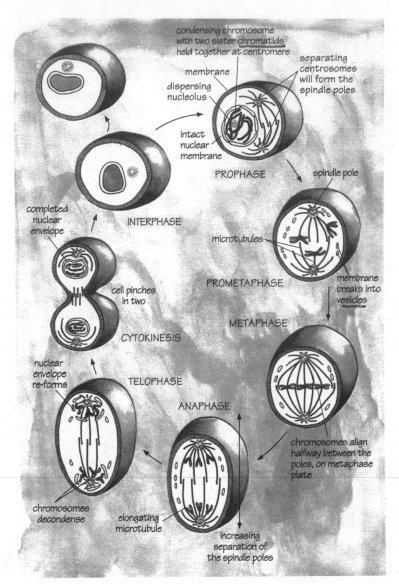

condensing chromosome
with two sister chromatids
held together at centromere

membrane

dispersing
nucleolus

separating
centrosomes
will form the
spindle poles

intact
nuclear
membrane

PROPHASE

spindle pole

completed
nuclear
envelope

INTERPHASE

microtubules

PROMETAPHASE

membrane
breaks into
vesicles

METAPHASE

cell pinches
in two

CYTOKINESIS

nuclear
envelope
re-forms

TELOPHASE

ANAPHASE

chromosomes align
halfway between the
poles, on metaphase
plate

chromosomes
decondense

elongating
microtubule

increasing
separation of
the spindle poles

**Figure 6.4** Cell division begins (top right) after chromosomes have been duplicated and condensed into the "fat X" form. The centrosomes (each containing two centrioles) move to opposite sides of the cell as the nuclear membrane breaks down. Microtubules (also called spindle fibers) grow out of the centrosomes and reach into the mass of paired chromosomes. By metaphase, each chromosome pair has microtubules attached to each member of the pair and all pairs align halfway between centrosomes. Then the chromosome pairs split, each member pulling itself toward one centrosome. At telophase the sorted-out chromosomes cluster together and new nuclear membranes from around each cluster. At the same time the cell elongates and begins to pinch off in the middle.

living-room cell these fibers are about as thick as clothesline rope. They are microtubules, the same kind of filament that forms a transportation network within cells. But here microtubules perform a different job, indicating once again that life often works by adapting one type of structure to a variety of purposes. The original use of the microtubule was probably in cell division (a function essential to life itself), and the other functions, such as tracks for motor molecules, probably arose later. The spindle fibers grow by the polymerization of small proteins called tubulins that, under the right conditions, spontaneously assemble into long, hollow tubes.

While the spindle fibers are growing out in all directions from the centrioles, making them look like two dandelions gone to seed, the chromosomes attain their most compact form. The simple secret of chromosome sorting is that the two members of each pair are still stuck together. It is as if the pairs of socks are tied together. Every time Sorter A pulls out one sock, its mate is automatically lifted into position for Sorter B to grab it and pull. Every random dip into the drawer automatically brings the matched other sock into position for a flawless separation. The cell's chromosome-separating task is accomplished in much the same way, the spindle fibers acting as sorters of chromosomes that are tied together by chemical bonds at a junction called the centromere. This is the place where the arms of the condensed chromosome pinch together.

Two specialized structures have developed at the junction, one on each side, facing in opposite directions (Figure 6.5). Called kinetochores, their job is to capture one or more microtubules coming from a centriole and hold on. If the microtubule is a rope thrown by a rescuer, the job of the kinetochore is to grab the rope so that the chromosome can haul itself to the safety of what will become a new nucleus in a new cell.

Of more than academic interest, the study of kinetochores may shed light on a sometimes fatal disease called scleroderma, the most visible symptom of which is a hardening of the skin. Victims, curiously, usually have antibodies that attack their own kinetochores. It is not known whether this leads to destruction of their chromosomes or failure of mitosis or some other problem.

Because kinetochores are always on opposite sides of the sister chromosomes, the act of a microtubule latching on to one side automatically orients the chromosome pair so that the other kinetochore faces the opposite centriole and its spindle fibers. Microscope studies of dividing cells reveal that the process of a centrosome putting out spindle fibers is highly dynamic. New fibers continually appear to be shooting out in random directions—actually they are polymerizing rapidly at the free end—and collapsing back—depolymerizing. If fibers from opposite poles happen to overlap and touch, they stabilize, apparently cross-linked by other proteins. They cannot link end to end because their polarities, or directionalities, are incompatible. Fibers also stabilize if they should happen to be captured by a kinetochore. Each kinetochore appears to be able to catch and hold half a dozen or so fibers.

Once all the kinetochores have grabbed microtubules, the opposing forces balance out and all the chromosomes are tugged until they line up on a plane halfway between the poles. Then the bonds linking the paired chromosomes break and the separated chromosomes move in opposite directions.

Movies of this process, taken through a powerful microscope developed by Shinya Inoue, show the chromosomes moving toward the poles, kinetochores in the lead and the "arms" of the chromosomes dragging behind.

It takes the kinetochores about fifteen minutes to pull the chromosomes from the middle of the nucleus to their new homes at either centriole.

For decades biologists had assumed that spindle fibers moved, pulling chromosomes toward the poles. The fibers were even called "traction fibers," implying that they pulled on the chromosomes. In the mid-1980s two cell biologists at the University of California at San Francisco reported findings that challenged this view. Their report illustrates how the observation of an unexpected, and seemingly impossible, phenomenon can lead to a new discovery. In experiments on cultured cells Tim Mitchison and Marc Kirschner found puzzling evidence that the microtubules were falling apart—or in chemical terms, depolymerizing—at the chromosome's kinetochore. This made no sense in terms of the old view, which held that the job of the spindle fiber microtubule was to pull the paired chromosomes apart and tug them toward the opposite poles. How could a rope dissolve at the very place it is attached to the thing it is supposed to pull?

Mitchison and Kirschner's report triggered a renewed search for the ultimate mechanisms of mitosis. In the late 1980s two other labs—Gary Gorbsky and Gary Borisy at the University of Wisconsin and R. Bruce Nicklas at Duke University—reported separate experiments that confirmed Mitchison and Kirschner and have laid the old belief to rest. Their data show that the spindle fibers stay still while the kinetochores do the real work, pulling the chromosomes along the fibers, something like a mountain climber working a rope hand over hand.

Nicklas's experiment, which was done on cells from grasshopper gonads (which he favors because they are unusually large and easy to manipulate), was simple in design. He put some of the cells on a slide under the microscope and looked for those that were dividing and in the stage where the two sets of chromosomes were traveling toward opposite poles. Then, using a microscopically fine knife (actually it was the edge of a very thin glass needle), he chopped off all the microtubules linking one set of chromosomes to one pole. He pulled the detached spindle pole well out of the cell, essentially discarding it. The cut ends of the microtubules that were left just laid there, going nowhere. There was nothing left to pull on them. But, as Nicklas watched in amazement, the chromosomes kept on moving toward the chopped-off ends of the fibers as if nothing happened. When they came to the ends of the microtubules, they stopped. Nicklas says his experiments also showed that the spindle fibers must be cross-linked one to another so that, even when severed from the poles, they are held in position while the chromosomes pull their way along them. If not, the much more massive chromosomes would sit still and simply pull the spindle fibers toward themselves.

"That doesn't happen," Nicklas says. "So that means something's gotta be holding the spindle fibers in place. We know there are a lot of proteins that bind to microtubules and they're probably holding them all in position."

As a result of such experiments, cell biologists can now accept Mitchison and Kirschner's curious finding that as a kinetochore moves toward the pole, its end of the microtubule depolymerizes.

"We have no idea how all this happens but you can think of it as the kinetochore sort of munching its way along the microtubule like a little Pac Man," says Wisconsin's Gorbsky, whose findings also confirmed the new view.

Taken together, the new findings focus major research attention on the kinetochore, an intracellular machine that most biologists had barely heard of a few years ago. It is now clear that the kinetochore is a crucial player in one of the most fundamental processes in the lives of non-bacterial cells. The search is on now to understand how it works.

"It's doing the pulling, so there's gotta be something like a motor molecule there," says Nicklas. "The question is, what kind of motor?" One candidate, introduced in Chapter 2, is dynein, one of the protein molecules that can flex at its own internal hinge. *Dynein* is the "uptown train" inside cells that carries vesicles of cargo in one direction along microtubules while *kinesin,* the "downtown train," carries them in the opposite direction. The two motors move in opposite directions because they somehow read a directionality inherent in the structure of microtubules. Dynein is a candidate because it can read microtubule polarity and move along it the right way to work in a kinetochore. Nicklas speculates that if dynein could bind one end of itself to part of the kinetochore, its other end could bind to a microtubule. When the dynein flexed, it could pull the chromosome and microtubule together. Then, as the motors reached down the microtubule to grab a new section, the part that had already been hauled in could depolymerize (Figure 6.5).

Errors in this chromosome-transporting mechanism are thought to be the ultimate cause of the most common form of Down's syndrome, according to B. R. Brinkley of Baylor College of Medicine in Houston, another specialist in kinetochores and mitosis. The disease is usually the result of a victim's cells each having three Chromosome 21s instead of the normal two. Biologists are puzzled as to how a mere extra dose of otherwise normal genes can be so harmful, but it is clear that the effect involves many other physical features influenced by at least several of the thousands of genes carried on Chromosome 21.

Down's syndrome begins before conception when the mother's ovary is making eggs. This happens in a chromosome-separating process called meiosis, which is similar to mitosis. Mitosis, the process we have just been discussing, separates duplicated pairs of chromosomes, giving each daughter cell the same full set. Meiosis, by contrast, separates unduplicated pairs of chromosomes and gives each daughter cell (sperms or, in this case, eggs) only half a set that contains one member of each pair. At conception, the normal double dose of Chromosome 21s is restored when the sperm contributes its one Chromosome 21. However, if a kinetochore in a cell of the mother's ovary is defective, the two chromosomes can fail to separate, producing an egg with two Chromosome 21s. If this egg is fertilized by a normal sperm carrying one copy of the chromosome, the embryo will have three copies of Chromosome 21 per cell.

As the separated chromosomes move toward their destinations, four new events happen. First, the poles themselves move farther apart, to opposite sides of the cell. How this happens is not entirely clear, but one theory is that the overlapping spindle fibers are pushed in opposite directions, possibly by motor molecules that slide each fiber past the other. And because microtubules are fairly rigid, the poles move apart. Also, part of the force may come from the spindle fibers that point away from the chromosomes and toward the cell's outer membrane. If those fibers reach into the cytoskeleton lining the inside of the cell membrane, a motor molecule there may pull on the fibers, drawing the two centrioles apart. Whatever the process, the newly separated clusters of chromosomes move to opposites sides of the cell.

Then the chromosomes begin decondensing, reverting to the thread-like, unraveled state in which they spend most of their time and in which their coded messages can be read. Also, new nuclear membranes begin forming over the surface of each chromosome, reassembled from the stored components of the old membrane. This is how the new nuclei keep out the organelles that are not supposed to be inside a nucleus. If the nuclear membrane formed as a full-sized ball, it would capture lots of extraneous molecules and organelles that could cause trouble. But because the new membrane starts as a skin tightly fitted to the chromosomes, the process ensures that nothing is inside except the chromosomes. The chromosomes are close together and the membranes quickly fuse. Gradually the receptors and other gatekeepers in the membrane allow access to those

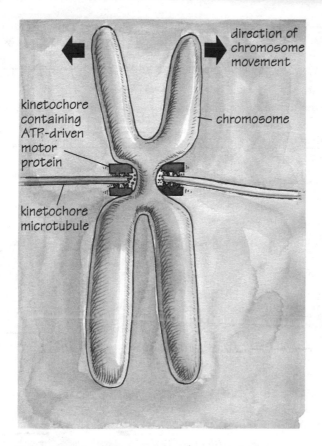

**Figure 6.5** At the wasp waist, where two chromosomes are linked, little-known structures called kinetochores (shown very dark here) are thought to pull on the microtubule, using motor molecules powered by ATP.

things that are supposed to get in. DNA polymerase, for example, carries a chemical address label that allows it to be transported back into the nucleus. As the chromosomes go back to their uncondensed, threadlike form, the nuclear membrane essentially inflates itself to normal size.

The final step in cell division is to pinch the mother cell in two. The cleavage plane cuts precisely where the twin chromosomes lined up before they separated. The cause of the motion is fairly well understood. A band of filaments, called the contractile ring, encircles the cell at the cleavage plane like a purse string. The filaments are made of the same two proteins whose interactions make muscles contract and that are thought to make cells crawl—actin and myosin. As we saw in Chapter 2, on cell motility, the head of the myosin molecule is a molecular motor capable of grabbing another molecule, pulling on it, letting go, and grabbing again. The myosins themselves form a parallel fiber by twining their long tails together. As the myosin filaments pull themselves along the actin filaments, the purse string tightens and squeezes the mother cell's membrane until it pinches the cell in two. Where there was one unit of life, there are now—approximately one hour after the chromosomes first condensed into the X form—two units of life. Only now do the chromosomes fully come out of suspended animation, unfurling into their working form.

Cell division is the process by which single-celled organisms reproduce. Within the republic of single cells that is a human body, all the same events are still the means of reproduction, turning the one cell formed at conception into the trillions of a fully developed human being. It is also the process by which previously quiescent cells proliferate to repair wounds and sometimes, unfortunately, to form tumors. We'll explore all these phenomena in later chapters.

It is obvious, however, that cell division cannot be allowed to happen willy-nilly. If it did, cancers would be sprouting throughout the body. Yet the cycle of events through which cells move—growing in size, replicating their chromosomes, dividing, and growing again—must roll on at a steady, controlled pace in certain tissues such as skin. In the culture dishes of cell biologists, many cell lines proliferate with mathematical precision. Every time a cell grows to double its original size, it splits in two. This is simple arithmetic. When a cell divides, its daughter cells are each half the volume of the parent cell. Then each daughter feeds and grows to full size. In a body that is growing, as in childhood, the growth of new cells adds to the body's size. And in many tissues, the pace of the cell cycle must be fairly fast, increasing the overall number of cells. But in an adult, the cell cycle must remain steady for tissues to stay the same size, new cells being created only to replace old ones that die. But the cell cycle must explode with activity when the body is wounded and rapid healing is needed. And then it must brake itself back under control when the healing is complete.

The frequency with which cells divide varies greatly. The fastest known cells, those of the fruit fly embryo, divide every eight minutes. They do not pause to grow or carry out any special functions before dividing again. Only when the requisite number of cells is generated do fly cells begin to do something else. Among the slowest are human liver cells, which can function for more than a year before receiving the signal to divide. Of course some cells, such as those of muscle and brain, never cycle once they reach adult proportions. Cells in a typical laboratory culture dish divide about once a day.

For more than a generation, cell biologists have searched for the mechanism that regulates the process—the controller that chooses the right time to start cell division and then orchestrates the events so that they occur in the right order. To understand that process, it has long seemed, would be to understand much about how life works and, of great practical value, how it goes awry in diseases such as cancer. Although major advances have come in recent years—some of the factors that determine the speed and frequency with which cells divide have been discovered, for example—much remains a mystery.

Cell biologists consider there to be four key stages of the cell-division cycle (Figure 6.6). The first starts just as new cells are born from the previous division. This is the period in which the cell lives the life for which it is intended, carrying out its biological functions within the body. To researchers who focus on the division process, however, this time is known simply as $G_1$ for "Gap 1." To those scientists, the real fun begins at the DNA-synthesis phase, or $S$, when the cell starts making a copy of its chromosomes. Then comes a second gap, $G_2$, which probably serves as a safety margin, ensuring that all the chromosomes are copied before the next fateful step, mitosis, or $M$. In about one hour, the nuclear membrane breaks down, duplicated chromosomes are separated, new nuclear membranes form around each of the two clusters of chromosomes, and the cell pinches in two.

Clearly, something must control the process. Cells don't pinch in two until after they duplicate their chromosomes. And once DNA replication occurs, the mechanism stops after exactly one copy is made. The nuclear membrane doesn't break down until the centrioles start sending spindle fibers toward the chromosomes, and it reassembles only after the chromosomes are segregated. And the chromosomes don't decondense to their functional threadlike form until the membrane has isolated them from the rest of the cell.

The basic sequence of events had been known for decades, but it was not until the 1970s that scientists found any of the molecules in the cell that regulated the process. In 1971 Yoshio Masui, then at Yale University, and L. Dennis Smith, then at Argonne National Laboratory, independently found a substance in frog eggs that when given to cultured frog cells caused them to divide. Although the exact nature of that substance would not be pinned down until the late 1980s, their findings showed that mitosis is triggered by the appearance of a molecular signal.

In the 1980s the search for what controls the cell cycle began to take major steps forward and is today one of cell biology's most exciting and most promising pursuits. The latest excitement began with a series of discoveries at the Marine Biological Laboratory in Woods Hole. Joan Ruderman, a cell biologist then at Duke University and now at Harvard, was teaching a summer course to graduate students. Tim Hunt, then at the University of Cambridge in England, was a summer investigator at MBL

**Figure 6.6** The cell cycle is often depicted this way to suggest the relative lengths of time a cell spends in each phase. After mitosis, M, the new cell enters the longest part of its life cycle, the misleadingly named first gap, or $G_1$. Though a time-out from cell division, this is the phase of a cell's life when it performs its chief duties for the organism. Cells that do not multiply, such as those of the muscle or nerve cells, remain perpetually in $G_1$. Others, however, enter S, which stands for synthesis of DNA (the phase in which chromosomes are duplicated). After a shorter gap, $G_2$, the cell actually divides in the process of mitosis, or M.

who collaborated with Ruderman. They worked primarily on sea urchins and the giant surf clams, both of which grow abundantly in the local waters. The urchins are the same species that early in this century brought Jacques Loeb such renown for artificially starting their embryonic development. The clam is the same species used for chowder. The females of both animals are ready sources of eggs that can easily be extracted and fertilized with sperm from the males. Both species readily proceed through the early stages of embryonic development in a laboratory dish.

During the early 1980s, Hunt and Ruderman, with the help of various graduate students at MBL for the summer, discovered that as the cells divided during embryonic development, they were producing certain proteins that behaved in a very odd way. First, their quantity grew during Gap 1, the interval between cell divisions. The proteins kept accumulating as the cell entered S phase, while it replicated its chromosomes. The concentration grew up to the moment when the paired chromosome duplicates were ready to pull themselves apart and head toward opposite poles. Then, curiously, all the mystery proteins broke down, disappearing completely from the cells.

Ruderman and Hunt had no idea what the proteins were doing in the cell, but because their behavior correlated with the cell-division cycle, they named their new proteins cyclins. Were the cyclins controlling the cell cycle, they wondered, or was the cell cycle controlling them? No one knew at first. The notion that they played a fundamental role, however, was bolstered as cyclins were eventually found in other sea creatures commonly available at Woods Hole, such as starfish. Eventually, it turned out that cyclins could be found in cells of all species, including humans. It is always a good bet in biology that if a protein shows up in a wide variety of species, it is involved not with some specialized activity of that cell type or a given species but with the basic workings of life itself. It is another reminder that all life-forms on Earth are descended from a common ancestor and all still share versions of a few fundamental mechanisms that arose billions of years ago.

In the late 1980s, other researchers working on yeast cells, including Paul Nurse of Britain's Imperial Cancer Research Fund, were discovering a family of other proteins that also seemed to play a role. They were working with an odd strain of yeast, one whose cells divided normally unless the temperature was a little too warm. Then all the yeast cells stopped dividing, all at the same point in the cell cycle—just after the cells had replicated their DNA but before the two sets were to move apart. Apparently, the signal needed to take the next step was missing or, at least, nonfunctional at the warmer temperature. Since other yeast strains were unaffected by the warmer temperature, the researchers speculated that the odd strain had a mutant form of some protein that was needed for it to move into the next phase of cell division. As often happens in biology, the discovery of a mutant form of organism provides an opportunity to track down a key element in a biological process.

To test the thought that heat inactivated some necessary signal, the researchers took various genes from normal yeasts, spliced them, one at a time, into the mutant yeast cells and checked to see if warmth still blocked the cell cycles. Eventually they found a gene that rescued the mutants, allowing them to keep dividing in warmer temperatures. Logic told the researchers that this gene's product must be the normal version of the protein that was defective in the mutant strain. The transplanted gene must have been making a good version of the protein that was defective in the mutants. The researchers called their molecule *cdc protein* for cell-division cycle protein. Like the cyclin of Ruderman and Hunt, they soon found that cdc was in cells of all higher species, including humans. In recent years many cdc proteins have been found, and it is now clear that, like the cyclins, cdc proteins are ubiquitous.

In 1989 Ruderman and her colleagues showed that cyclins and cdc proteins do not work alone; they must collaborate to govern the cell cycle. Moreover, when the two proteins combine, they turn out to form the substance that Masui and Smith had discovered in frog eggs in 1971. In the early 1990s, researchers in many laboratories were slowly elucidating the way the two molecules work together. The specifics are complex and not well understood, but the basic idea is fairly simple. The cdc protein exists throughout the cell cycle, but by itself it cannot do anything. Cyclins, however, are repeatedly manufactured and then broken down over the course of the cell cycle. During $G_1$, for example, the cell makes a form of cyclin that binds to cdc, slowly building up the concentration of combined but still inactive molecules. The combined molecules are activated by a process that starts slowly but then suddenly surges. When the concentration of active combined molecules reaches a critical level, this triggers the nucleus to start replicating its DNA.

Once DNA replication starts, the cyclin part of the complex breaks down. Then during $G_2$, a second form of cyclin is manufactured and a similar process of buildup and activation triggers mitosis, the actual dividing of the cell. Then that cyclin is also broken down. What the molecular complex does to trigger each step is to attach a phosphate group (one phosphorus atom plus three oxygen atoms) to some other molecule. The process is called phosphorylation, and it is a common way that processes are regulated throughout the cell.

Most proteins have certain amino acids on their surfaces that are capable of binding chemically to a phosphate group. But when this happens, the phosphate adds its shape and chemical properties to that of the protein, creating a new molecule with different propensities to interact with other molecules. The phosphate may also cause the protein to change shape, adding a second alteration of chemical properties. When the phosphate is removed, the protein changes back. Many of the proteins that play various roles in life—quite aside from controlling the cell cycle—achieve their active form by being either phosphorylated or dephosphorylated. In other words, to make a given enzyme active, it may have to have a phosphate stuck on at a certain place. To switch off the enzyme, the phosphate must be removed. Other enzymes work the opposite way. They are held in an inactive form while phosphorylated. Another protein that takes off the phosphates automatically has the power to regulate the function of the phosphorylated protein.

The molecular steps that must be accomplished in precise sequence to carry out cell division, it now appears, are switched on and off by a series of phosphorylations and dephosphorylations of different proteins with roles to play in cell division. The particular protein to be phosphorylated at a given moment is dictated by the particular link-up of a given cyclin and a given cdc protein.

Although much progress has been made toward understanding the cell cycle, many unknown steps remain to be discovered. For one thing, it is not clear what the phosphorylation actually does to trigger the next step in the cycle. Also, it is not clear how cells know that each step has been completed before they start the next. The whole process would seem to need some feedback mechanism, some way that the cycle controller knows it is time to initiate the next phase.

What has been learned about governing the cell cycle in the last decade is enormous but it is clear that much more needs to be discovered to understand in detail the ability of a cell to reproduce itself—surely the most characteristic and the most astonishing of life's unique properties. Moreover, the proliferation of cells is also under the control of factors—presumably genetic—that can shut the entire process off, leaving certain kinds of cells indivisible for years, even decades. And quiescent cells can be triggered—usually by molecular signals from outside the cell—to resume furious cycles of proliferation, sometimes for good purposes such as to heal a wound, sometimes for bad such as to grow a tumor. Each of these phenomena will be dealt with at length in a later chapter.

For the human species, however, the first responsibility of cell division is to construct, from one founding cell, a person.

The creation of a new human being neither begins at "conception" nor ends at birth.

Although often obscured by the emotionalism of the abortion debate, these facts are becoming inescapably obvious as biomedical researchers battle against unwanted pregnancy and against birth defects. Scientists searching for better contraceptives, for example, are finding that the creation of a human embryo is not an instantaneous event but a process of many steps over a period of more than a week and, depending on how the embryo is defined, up to two weeks. Each step is, at least theoretically, a moment when some form of intervention could prevent pregnancy.

Many researchers say this realization fits well with the emerging view that the origin of any one human being is a process of many phases, not one of sudden transformation. None of this discussion, it should be clear, deals with religious beliefs or with any legal declarations as to when the developing embryo acquires rights or what rights they might be. Religion and law lie outside the realm of science.

From the time a man's sperm meets a woman's egg (also called an ovum or oocyte), there are several dozen necessary events and processes that must take place over a period of nine to ten days before the true embryo—the cells that will become the baby—can even begin to develop. As will become clearer in this chapter, popular usage of the term "embryo" is far looser than the scientific usage. Among developmental biologists, the specialists who study this process, "embryo" refers only to the mass of cells that have begun to differentiate into what will become the baby. During the first two weeks after fertilization, those cells remain undifferentiated while other cells in the "conceptus," or "pre-embryo," work on the higher priority task of constructing a placenta.

Researchers striving to prevent birth defects in the embryo or fetus (a stage of development, that comes about eight weeks after fertilization) are learning something similar but more dismaying: There is an incredibly complex series of steps in human development, and any of scores of possible missteps can prevent an embryo from arising or can kill the embryo before the woman ever imagines she is pregnant. Many of those missteps, however, can lead to the even more unfortunate result of a baby being born with profound birth defects.

As one waggish scientist summarizes the knowledge: "You can, in fact, be a little bit pregnant." It is generally accepted that on more than half the 1. occasions when a sperm fertilizes an ovum and development begins, the woman never learns she has become pregnant because the process fails for natural reasons before the next menstrual period is expected. Still more pregnancies, of course, abort naturally after a woman knows she is preg- 2. nant.

From research in many laboratories across the field, a remarkably intimate picture is emerging of how a human being is created, a picture that indicates the often-repeated statement "life begins at conception" is misleading on two counts.

First, life began once billions of years ago and ever since has simply been relayed from one cell to the next, from one generation to the next. The unfertilized egg is a living cell and unquestionably possesses life. The sperm, though highly specialized in shape and function, is also a living cell. Life exists in each one of the millions of sperms carried in a man's semen. When sperm and egg join in the process called *fertilization,* the resulting zygote has life but is not any more alive than were the sperm and egg. To be sure, however, fertilization creates something dramatically new and different—a new kind of living entity that has the potential to develop into a human being. Thus science can say that the key legal and moral question in the abortion debate should not be phrased: "When does life begin?" It should be put: "When does a life begin?" or, better, "When does a person begin?"

The second reason the "life begins at conception" view is misleading is that conception, as we shall see, is a long, multistage process, not an instantaneous or even brief event. For example, the moment when the sperm enters the egg occurs a full twenty-four hours before the two sets of genes come together to create the new genetic potential.

Moreover, any scientifically based definition of "conception" must also take account of the fact that many more events must happen over yet another week before any structure arises that will be part of the true embryo. The cells up to this stage are only laying the foundation for the start of the embryo by constructing a rudimentary placenta to obtain nourishment from the woman's uterus.

All of the events that lead up to the creation of the true embryo are now understood to unfold during a lengthy process. Even the scientific word "fertilization" is only a small improvement over "conception." It refers specifically to the act of the sperm penetrating an egg and the genes of the two joining. But this happens in a series of steps that takes about twenty-four hours. No matter whether you say that personhood begins with conception or fertilization, it is a gradual process, not a point in time.

Once every menstrual cycle, an egg bursts out of one of a woman's two ovaries (the ovaries take turns every other month—the result of a regulatory process still not understood). Caressing the ovary are the undulating "fingers" fringing the open end of the adjacent fallopian tube, also called the uterine tube or oviduct. The fingers move across the ovary and literally sweep the egg into the tube.

The egg at this stage is a full-fledged cell, with all the standard organelles, including—contrary to a widespread misconception—the human cell's normal complement of forty-six chromosomes arranged as twenty-three pairs of identical chromosomes. The human egg, though a thousand times bigger than most cells (it is about as big as the dot on this *i*), does not travel alone. It is escorted into the fallopian tube by thousands of tiny cells enclosing it in a "cumulus." The egg, or perhaps its escort cells, secrete special molecules to catch the attention of sperm cells, should any be present, and this substance diffuses through the fluid in the fallopian tube.

Nestled within its retinue of escorts and preceded by a chemical signal that might as well be called perfume, the egg glides through the tube toward the uterus, transported by the cells lining the tube. These cells have hairlike cilia projecting into the tube, and the cilia lash back and forth in coordinated waves. The egg in its cumulus virtually surfs the waves toward the uterus—a slow process that takes about four days.

If sperm have entered the woman's reproductive tract, they sense the egg's perfume and swim toward it by the millions, furiously lashing their tails to swim against the ciliary waves. Like the egg, the sperm is a living cell. But unlike the gigantic egg, the sperm is one of the smallest human cells (not counting its very long tail) and it is stripped for action. Its sole goal in life is to swim to an egg and deliver a packet of dormant genes. Sperm genes do not express themselves (unless they are lucky enough to fertilize an egg), so the sperm cell makes no messenger RNA, contains no ribosomes to read the mRNA, has no endoplasmic reticulum (ER) to help the proteins fold properly, and lacks any Golgi apparatus to process proteins for export.

What a sperm does have, and in great abundance, are mitochondria, the organelles that make ATP to supply energy to the cell. And to fulfill its goal, the sperm must race a distance several thousand times its own length (counting its very long tail), so it requires plenty of energy. On the scale of the living-room cell, the sperm is about five times as long as the living room (but its head is no fatter than an overstuffed chair) and it must swim nearly 2,000 miles. To supply the energy conveniently, all the many mitochondria are wrapped snugly around the base of the tail, as close as they can be to the site of the sliding microtubules that make the long tail lash back and forth. The sliding motion is caused by dynein motor molecules grabbing one microtubule and walking (or perhaps running) along the adjacent microtubule.

In possessing this long, whiplike tail, or flagellum, human sperm cells recall most vividly their evolutionary link to many kinds of free-swimming, one-celled protozoans. Moreover, sperm tails contain the standard arrangement of microtubules found in all cilia and flagella—two separate microtubules down the center, surrounded by nine microtubule doublets (which are pairs of microtubules fused lengthwise). So standard is this "nine-plus-two" arrangement that it is possessed by all organisms on Earth, from bacteria to humans, that have moving hairlike projections. Even plant sperm have the same arrangement. (Plants may be sedentary but their sperm swim as well as animal sperm.) The nine-plus-two arrangement is another powerful piece of evidence, as if anybody still needed it, that all life on Earth works according to a common set of structures and processes.

As it happens, one cause of male infertility (Kartagener's syndrome) is a defect in a man's dynein molecules, stemming from a defective gene. If the dynein molecule lacks the portion that extracts energy from ATP, it can't flex, which means the microtubules won't slide, which means the sperm can't swim. Of course, since dynein is also needed to make the microtubules in cilia slide, the ciliated cells lining victims' air passages are also immobile, unable to sweep inhaled dust and chemicals up out of the lungs. As a result, victims suffer from chronic bronchitis and frequent sinus infections. Before the advent of modern cell and molecular biology, no physician could have understood how infertility and sinus problems could result from the same cause.

The sperm armada, drawn by the egg's perfume, swims from the vagina through the opening in the cervix into the uterus and out the far side of the uterus into the fallopian tube. Typically between 300 million and 500 million sperm cells begin the quest. But a woman's reproductive tract can be a hostile environment, and most sperm die on the way. Ironically, unless sperm cells expose themselves to this environment for several hours, they are physically unable to fertilize an egg. This is because sperm cells emerge from the man encumbered with a coating—made of linked sugars and proteins—that must be stripped off. Apparently enzymes in the uterus or fallopian tube perform this service.

Only a few hundred sperm cells survive to meet the egg, usually somewhere in the tube. This high mortality rate is a common reason for male infertility. If the man produces too few cells in each ejaculate, there simply won't be enough that live to reach the egg and carry out the next steps. The first sperm to meet the egg, however, is not likely to fertilize it. The cumulus of escort cells is in the way and must be removed.

At the tip of each sperm is a chemical "warhead" that bursts open as it meets the egg, releasing enzymes that break down the "glue" holding the cumulus cells together. As more sperm cells batter the cumulus and release their enzymes, the little escort cells come loose, exposing the egg. The sperm cells also undergo further changes. The previous wavelike beating of their tails now changes to stronger, whiplike lurches, and the sperm plasma membrane is altered in several ways that are not fully understood.

But, still, fertilization is not a foregone conclusion. Jacketing the egg is a second protective barrier, a jellylike coating called the *zona pellucida*. Within this coating are embedded thousands of "recognition factors," molecules that are like receptors. They can bind only to other molecules of a complementary shape, and unless that happens, no sperm can fertilize the egg. The job of the recognition factor is to prevent the sperm of the wrong species from fertilizing the egg. The shape it is "looking for" exists only on proteins on the surface of sperms of the same species. The mechanism is a holdover from the time in the evolutionary past when human ancestors fertilized their eggs outside the body, in the open sea as many marine organisms still do. By granting admission only to sperms of the right species, this mechanism prevents the wasting of unfertilized eggs drifting in the open sea. If the egg's receptor fits the sperm's, the key opens the lock and the sperm is allowed to penetrate the zona (Figures 7.1 and 7.2).

Several laboratories are studying human recognition factors as potential avenues of contraception. One theoretical approach would be to vaccinate the woman with human sperm recognition factors. Experiments at the University of Connecticut, for example, have shown that in guinea pigs a vaccinated female develops antibodies that bind to the receptors on the sperm, blocking their access to the egg's recognition factors and preventing fertilization. The method also worked if males were given the same vaccine. They simply produced sperm whose recognition factors were already occupied by mimics of the egg's recognition factor.

Once a molecularly compatible sperm has forced its way through the jellylike zona, its outer membrane fuses with that of the egg. Here again sperm and egg receptors are thought to be involved. Once the membranes fuse, the sperm's internal parts begin to emerge like a snake from a shed skin. The sperm does not swim into the egg. Contractile filaments from the egg attach to the sperm and pull it in, tail and all, leaving its empty membrane, like a ghost, outside. But fertilization is far from over. The sperm head, holding the father's chromosomes, must wait a full day. The egg has much work to do first.

The egg's first priority at this point is to shut the door as soon as one sperm gets in. If a second sperm got through, its extra dose of genes would throw the fertilized egg into a lethal confusion. Eggs fertilized by more than one sperm soon die. To prevent this, the egg relies on its calcium

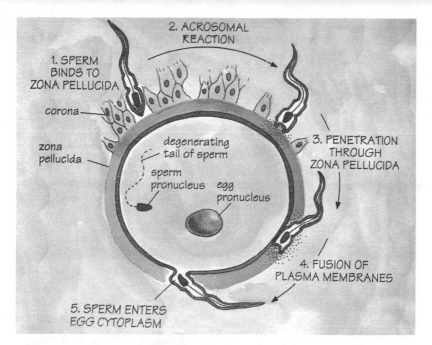

**Figure 7.1** Fertilization starts as sperm batter the small corona cells surrounding the egg, releasing the enzymes in their acrosomes to break down the coronal bonds. If special "recognition factor" proteins in the egg's jellylike zona pellucida recognize and bind to complementary proteins in a sperm, it is allowed to penetrate the egg's membrane, where other proteins must interact to cause membranes of egg and sperm to fuse. This delivers the sperm's nucleus and tail inside the egg.

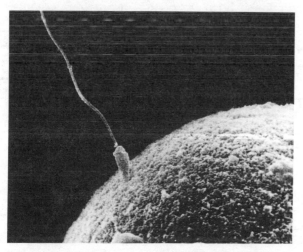

**Figure 7.2** A microscope's view of a sperm approaching an egg. (Photo courtesy of American Society for Cell Biology)

channels, gateways that can quickly import calcium atoms into the cell. As soon as a sperm gets in, a surge of calcium atoms, each carrying a small positive electrical charge, flood the egg, changing the charge on its surface from negative to positive. This creates an electrostatic field that repels other sperms.

The electrical charge also triggers the egg's program of development, starting a sequence of cell divisions intended to multiply the number of cells in the early conceptus. As it happens, this trigger can be pulled without any sperms at all, and cell biologists routinely exploit this effect to make egg cells divide in the laboratory. The experiment is most often done with nonhuman egg cells. The sea urchin is a favorite because it is easy to trick the female into shedding thousands of eggs at once. A mild dose of potassium chloride is poured into a test tube, and because of the electrical charge it carries, it pulls the egg's developmental trigger. Within minutes all the eggs in the dish behave as if they had just been fertilized by sperm. Each single cell divides into two, then two into four, and so on. At no time in these early divisions is the sperm needed. All the instructions and apparatus for carrying out the first few rounds of cell division are provided by the egg, presumably in the form of enzymes formed while it was in the woman's body and perhaps even in the form of messenger RNA waiting to have its message translated into protein upon some signal received during fertilization. Whatever the mechanisms, the human egg appears to be fully equipped to carry out all of the work and the guidance needed for the first several rounds of cell division with no help from the sperm.

The first role of the electrical charge—to prevent fertilization by more than one sperm—is brief. It lasts for only a few seconds—just long enough to erect a more permanent barrier. The egg releases a substance into the zona. Like an epoxy hardener, it makes the once gelatinous zona tough and impenetrable.

Still, however, the genetic union of sperm and egg must wait. The sperm has twenty-three chromosomes, one representative of each pair in the father. But the egg still has forty-six chromosomes (both members of each pair in the mother), double the correct amount for fertilization.

While the sperm takes a break—it kills time by sitting just inside the egg's outer membrane—the egg's twenty-three pairs of chromosomes split apart. Each set of chromosomes goes into one of two new nuclei. Then the egg divides but in an unequal fashion. It pinches in two to make a tiny cell that holds one nucleus and a giant cell that holds the other nucleus. The tiny cell eventually disintegrates. In the meantime, the sperm's head, which had carried the father's chromosomes in a supercondensed and highly compact state (the same state they assume during mitosis), expands to many times its original size and its tail disintegrates.

Biologists long thought that at this point the sperm's nucleus with its twenty-three chromosomes finally joined with that of the egg and its twenty-three chromosomes to make a single cell with the normal complement of forty-six. In recent years, it has been discovered that this doesn't happen. Instead each nucleus duplicates its twenty-three chromosomes independently to make forty-six each. Then, at last, the two nuclei begin the final phase of their merger. The nuclei move toward each other, presumably pulled by the egg's cytoskeletal filaments and some unknown motor protein. The membranes around each nucleus break down and all ninety-two chromosomes assemble on a single plane, just as if an ordinary cell were undergoing mitosis. Microtubules reach in from opposite poles of the egg and attach to opposite sides of the chromosome pairs. The chromosomes, through the apparatus at their kinetochores, pull themselves apart and head for opposite poles of the cell. A belt of actin fibers encircling the egg cinches down, pulling the cell membrane with it to pinch off between the two poles. The product is two genetically identical cells, each with a half-set of chromosomes from each parent. It has taken thirty hours after the sperm entered the egg to reach this stage, the first point at which a new genetic potential has been established. Nothing has been done yet specifically, however, to create an embryo. There is much more work to be done before that task can be started.

For the next few days, the conceptus continues its leisurely journey toward the uterus. Had it not been fertilized, it would have lived only ten to fifteen hours after ovulation. (Since sperm can survive for about forty-eight hours after ejaculation into the vagina, for pregnancy to occur, sexual intercourse must happen within a narrow window that begins forty-eight hours before ovulation and ends ten to fifteen hours after.) Eggs that are never fertilized are eaten by macrophages that normally roam the inner surface of the fallopian tube. These cellular undertakers also feast on dead and dying sperms that never find an egg.

During Day 2 the two cells become four, all identical. Then four become eight, all still identical. Each division cuts the size of each cell in half. The conceptus, still imprisoned inside the shell of its hardened zona, cannot feed or grow in mass, although it converts much of its stored nutrients into the new DNA needed for each new set of chromosomes and into the other organelles that are multiplied. The conceptus at this stage is called a *morula* because it looks something like a mulberry and "morula" is Latin for the fruit (Figure 7.3).

This, incidentally, is the stage at which the "test tube baby" conceptus, produced by fertilizing the egg in a dish or test tube, is injected into the prospective mother's uterus. More than 10,000 babies have been born using this procedure, usually called in vitro (Latin for "in glass") fertilization, or IVF. To perform IVF technicians must themselves carry out some of the steps of fertilization that happen normally in the uterine tube.

New research is showing that IVF is an attractive option not just for

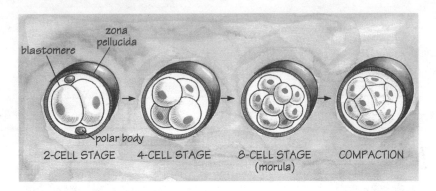

**Figure 7.3** After fertilization, the next few rounds of cell division take place within the hardened zona pellucida. The cells cannot feed or grow, so each division cuts the size of each in half. The last transition shown here is the first formation of links among the cells. These are bonds and junctions by which each cell pulls itself close against its neighbors.

infertile couples but for those who risk having a baby with a genetic disease inherited from one or both parents. In the past, such couples have either abandoned any hope of being "biological parents" or have resorted to prenatal diagnosis (usually using amniocentesis); if the embryo appeared to have inherited the defective genes, abortion would be an option. IVF offers the opportunity to diagnose the problem before the embryo can implant itself in the uterine wall and, therefore, to avoid the need for abortion. Embryos in a test tube can be diagnosed by removing one cell from the eight-cell stage and testing its genes. At this stage, the cells are completely undifferentiated and removing one has no effect on normal development. If the cell harbors a defective gene, the embryo can be rejected. Since IVF usually produces several embryos at a time, technicians can go on to the next one and test it. If it looks good, it can be implanted.

At the eight-cell stage, the cells of the morula undergo an epochal transition. Up to this point they have been running on autopilot, completely inert, their genes issuing no instructions. All the events up to this point have been guided by enzymes already present in the unfertilized egg and by messenger RNA previously transcribed from the now-dormant maternal genes. Enzymes and mRNA were waiting inside the egg when it left the ovary two or three days earlier with a set of plans to be carried out only if a sperm fertilized the egg and after the zygote underwent the first three rounds of cell division.

Only now, three days after fertilization, does the new combination of genes begin to express a new genetic individuality. At this stage, the conceptus begins to take control of its destiny. And at this point a great many morulas die because they are carrying catastrophic genetic defects that are only now expressed. Some experts estimate that as many as half of all fertilized eggs succumb at this point.

One of the first effects of the newly awakened genes is to direct the cells to cooperate in the creation of a structure that is greater than any one cell. Individual units of life, individual cells, band together into a rudimentary society of cells and form a very simple kind of tissue. When viewed under a microscope, the cells appear to squeeze together more tightly. Before it happens the conceptus looks like eight balls, stacked four on top of four. Afterward, the cells are one large, lumpy ball—the mulberry look. Embryologists call the event *compaction.* What actually happens is that the cells begin making a certain kind of specialized receptor molecule and installing it in their outer membranes. The molecules are receptors for one another, causing each cell to stick to its neighbor. As more receptors enter the membrane, the area of adhesion between cells grows, pulling them into the compact morula shape.

The cells also begin linking together in other ways, forming four kinds of junctions. These are the beginning of a communication system that human cells will use to coordinate their activities for the rest of the life of the individual. Some junctions simply bind one cell to another—far more tightly than the adhesive receptors. Other junctions provide channels through which small molecules may pass between cells. Over the life of the individual, networks of such junctions will be created, modified, and destroyed many times over, as the cells reorganize themselves again and again into a changing panoply of tissues and special structures intended to perform special roles.

One of the binding junctions is often thought of as a spot weld holding two cells together. On the scale of the living room cell it would be about the size of a poker chip and would look like one stuck against the wall. If the poker chip were a spot weld holding your living room to the room next door, there would be a similar spot on the corresponding point in the other room's wall. Connecting the two poker chips—in the space between the cells, which would be about a quarter of an inch—would be a web of tough filaments (Figure 7.4). Also, from the inside face of each poker chip there would be a number of string-sized tough filaments reaching into the room. In fact, the filaments from a spot weld on one side of the room might cross the entire room and link with another spot weld on the far side. These junctions, called spot desmosomes, provide much of the strength that holds cells of a given type together in a tissue. Many cells in the body are also joined by belt desmosomes, similar junctions where tough filaments link each cell to its neighbor. Instead of having strings that hang across the living room, they have filaments that lie flat against the wall and which may go all the way around it. But in the eight-celled morula, the only tissue is made of the eight cells themselves.

The eight cells now begin to form two other kinds of junctions. There are structures that under the microscope look like more welds, but unlike the spot welds, they appear to have no filaments attached to them. Such junctions are well known among cells that must join in the adult body to form a sheet or vessel that keeps fluids on one side from slipping between the cells to leak out the other side. For example, cells lining blood vessels are like this, as are cells lining the gut. These are called *tight junctions,* but little is known about how they work.

Morula cells also form much smaller structures called *gap junctions,* which are tubes that extend through the membrane of one cell to fit similar tubes extending from the neighboring cell. The result is a small passageway linking the cytoplasm of one cell with that of its neighbor (Figure 7.4). A typical cell would have many thousands of such junctions to its cellular neighbors. The tubes, which are of course made of proteins, cross the narrow gap between cells and let through small molecules—up to the size of sand grains in the living-room cell. This means that water can pass easily, as can certain small molecules that act as messengers. For example, if a receptor on one cell receives a hormone (sometimes called "first messenger"), it may release a different molecule (the second messenger) that acts on the cell to trigger various responses. Because second-messenger molecules can go through gap junctions, the hormone's effect may actually spread to neighboring cells. In this way a small signal may produce a large biological effect.

All of these junctions are vitally important to the functioning of virtually all cells in the human body, where their roles are much better understood than in the morula. Nonetheless, their appearance at this stage of development indicates that the eight cells are beginning to cooperate in the first steps of a very large task. In a sense, they are beginning to share information.

In the fourth doubling of cells, on Day 3 after fertilization, another epochal event happens. For the first time, the cells display different properties and begin to pursue different fates. Some of the cells will take one developmental road while others go a different way. If the cells of an embryo were unable to strike off like this and develop in directions different from those of their neighbor cells, there could be no development. No multicelled organism would be possible. If cells just kept on dividing but always kept the same forms and functions, they would create nothing more than a great amorphous lump of homogeneous tissue. Actually, the lump couldn't get very big because once a given cell found itself surrounded by other cells and shut off from a source of oxygen, it would suffocate. Most human cells must stay within a few cell thicknesses of a source of oxygen or they will die.

So how do the homogeneous cells of the early conceptus become different and how do they know what kinds of differences to express? Why

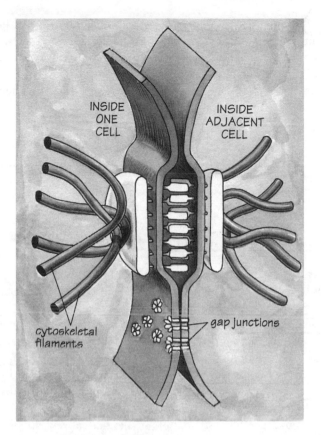

**Figure 7.4** A schematic view of the structural relationships between two adjacent cells. Gap junctions allow small molecules to pass back and forth. Larger structures are spot desmosomes, points on the cell membrane where the cytoskeletal filaments of one cell are strongly anchored to those of an adjoining cell. This is what gives many tissues resistance to falling apart or even to being pulled apart.

don't skin cells secrete digestive juices? Why don't cells of the eye's retina crawl away to eat invading bacteria? Every cell of the body carries exactly the same full complement of genes. But each cell expresses only a tiny portion of its genetic library. Somewhere in its history, somehow each of the 60 trillion cells of the republic of the human body has accepted a highly specialized role and forsaken the use of all the rest of its biological potential. How this happens, most biologists agree, is one of the central problems of modern biology. It is only beginning to be understood.

The first instance of cellular differentiation in human development—which arises during the fourth doubling of cells, from eight cells to sixteen on Day 4—is now beginning to be understood. Because experiments on living human embryos are rarely done, much of the research on this stage of development is based on experiments with mouse embryos. As mammals, their early development is believed to be much like that of a human, if not identical to it. In fact, it is generally accepted that the phenomena and mechanisms of development should be the same. Mice and men possess the same array of tissues and cell types.

Shortly after the eight-cell conceptus undergoes compaction, the outer region of each cell becomes very different from the rest of the cell. The surface develops hundreds of tiny, fingerlike projections that reach out into the surrounding medium. They touch the zona pellucida, which still encapsulates the conceptus. Just inside the surface of the cells a dense network of cytoskeleton forms, made of microtubules and actin filaments. Many organelles become concentrated just under this cytoskeleton. By contrast, the other end of the cell, toward the center of the eight-cell ball, develops no special features and is relatively devoid of organelles. It is as if each cell were concentrating its resources against its outer surface.

Then the eight cells divide in unison but the cleavage plane is not in the same orientation in all eight. Two kinds of daughter cells are formed— those that acquire some or all of the structurally enriched outer surface and those that acquire none of it. On average, about eleven of the sixteen daughter cells inherit a piece of the outer surface and they remain in the outer layer of cells. But about five cells wind up with none of the outer surface and they find themselves in the center of the ball, each equipped with a nucleus and a few essential organelles but entirely surrounded by the eleven surface cells rich in cytoskeleton and other structures. There are now two very different cellular environments, and the two populations of cells begin to behave differently. The conceptus has undergone its first differentiation.

The outer cells, called the trophectoderm, will go on to develop into a more complex apparatus for invading the wall of the mother's uterus while the little clump of inner cells multiplies slowly without developing any specialized features. These cells, which embryologists call simply the inner cell mass, are the ones that will, if the outer cells can invade the uterine wall, develop into the true embryo. Japanese researchers working on mouse embryos have found evidence that development of the inner cell mass into an embryo is actually inhibited by chemical signals from the outer cells. Using microsurgical instruments, they plucked cells of the inner cell mass out of the surrounding cells and found that they quickly changed to acquire the form of the surrounding cells, complete with the fingerlike projections on the outer membrane. Somehow signals from the outer cells must be blocking activation of certain genes of the inner cell mass.

At about this sixteen-cell stage, the morula reaches the uterus but drifts about unattached while it undergoes its next major change. During this time, its cells double again and again. Also, it absorbs oxygen and nutrients from the uterine fluid. The outer cells open certain junctions so that fluid from themselves, and perhaps from the uterine cavity, can flow into hollow spaces between cells in the middle of the ball. The outer cells flatten as the fluid inflates the ball. The inner cell mass does not participate in this step and remains a humble clump of undifferentiated cells stuck against one inner wall of the ball, bathed in the incoming fluid. The conceptus is now called a blastocyst.

Starting around Day 4, an extraordinary thing happens only to female embryos. One of the chromosomes in each cell is put out of commission permanently. It is one of the two sex-determining chromosomes designated X. All embryos inherit an X from their mothers and either a Y or an X from their fathers. If the resulting combination is XY, the embryo is male. XX is female. In male embryos both sex chromosomes function throughout life. In females, however, one is permanently inactivated. It shrivels up and never expresses its genes. As a result, human females have one less working chromosome than do males.

The choice of X chromosome to be inactivated is random in each cell. Once the choice is made, however, that same X chromosome remains inactive in all cellular descendants of the original. Somehow, through a process that remains unknown, the shriveled X chromosome, though replicated at each round of cell division, shrivels up again in each daughter cell. Only in one part of the body will the inactivated X chromosome return to normal form—the parts of the ovaries where new eggs are made. Half the eggs will have working copies of one X while the other half have working copies of the other X.

X inactivation means women are genetic mosaics—some parts of a woman's body are derived from cells with an active X chromosome from the father while other parts of her body have the mother's X chromosome. The effects of this are unknown, but since the mother and father carry different X chromosomes, the exact set of genes expressed in the next generation will depend on which cells are expressing which X chromosome. This leads to an anatomical mosaic in which, say, skin in one part of the body behaves one way while skin elsewhere may behave another.

The consequences of this are unknown, with one possible exception. It may explain why a few women come down with a form of muscular dystrophy, a disease that overwhelmingly prefers males. This is because the gene that fails is on the X chromosome. Women have two, so there is usually a good copy of the gene on one X that can make up for the loss of the equivalent gene on the other X. Males, of course, have only one X, so if that one is bad, there is no alternative. But if a female inherits the defective gene on one of her Xs and the other X is inactivated and if the clonal descendants of that cell happen to make up a significant proportion of a muscle, that muscle can become dystrophic.

Also on Day 4 or Day 5 the blastocyst must "hatch," as embryologists say, out of the zona. The job falls to the outer cells, which manufacture and secrete an enzyme that breaks down the zona. (Incidentally, when this was discovered, it suggested there might be a possible new method of birth control. If it were possible to block this enzyme, the blastocyst could never hatch and never implant. The idea is speculation, but it illustrates how a detailed knowledge of these phenomena is leading to new ideas with practical applications.) Freed from the zona, the outer cells take on another special job. They manufacture and release a hormone into the uterine tube, a substance called chorionic gonadotropin (CG) whose job is to get to the uterine lining ahead of the blastocyst and tell it to prepare. (Here again is another idea for a contraceptive. Some researchers think that if a woman could be vaccinated with CG, she would make antibodies that bind to it and render it unable to deliver its message to the uterine lining. When the conceptus arrived, it would be unable to implant and would disintegrate.)

After hatching, the six-day-old blastocyst prepares to invade the uterine wall. Still no bigger than the original ovum but now made up of about 120 cells, it produces special enzymes for the task. It digests away the lining of the uterus and embeds itself in the maternal tissue (Figure 7.5).

This is one of the few times in the human life cycle that cells make enzymes that can digest holes in the lining of other human tissues and then travel through them. Cancer is another time. When a tumor turns malignant, its cells acquire the ability to eat their way out of the tissue where they arose and spread to other parts of the body. Some specialists suspect that cancer cells somehow reactivate the enzyme genes that were supposed to stay dormant after implantation.

On Day 7, as the blastocyst is digesting its way into the uterine wall, some cells from the outside layer of the hollow ball multiply but without forming new membranes to separate them. The result is a massive single cell with many nuclei that envelops the blastocyst. This bizarre structure sprouts tentacles, which are presumably powered by the same kind of actin polymerization that enables motile cells to crawl. The tentacles push deep into the lining of the uterus. Over the next two days, as the blastocyst chews its way deeper into the uterine wall, the tentacles grasp the mother's blood vessels. Enzymes from the tentacles break down maternal tissue, rupturing blood vessels and spilling oxygen-rich blood into the empty spaces in what is becoming a spongy, blood-engorged tissue—the primitive placenta. The blastocyst's tentacles extract oxygen and nutrients and relay them to the inner cell mass.

The outer cells of the blastocyst keep up their output of CG, which commands the mother's body to suspend the normal menstrual cycle—

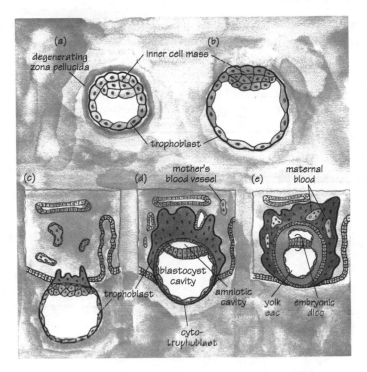

**Figure 7.5** (a) The blastocyst's zona pellucida breaks down about the time it reaches the uterus and prepares to implant. (b) Also, the first hint of the part that will become the true embryo appears, a cluster of a few cells called the inner cell mass. The rest of the blastocyst is the outer shell of cells, or trophoblast, which is destined to become the placenta. (c) Trophoblast cells produce enzymes that break down the uterine wall. (d, e) Parts of the trophoblast invade deeper and embrace maternal blood vessels to capture nutrients and oxygen needed for further development.

which would otherwise slough off the uterine lining in a few days. After CG enters the mother's bloodstream, it reaches the now-empty follicle (which released the ovum) and triggers it to develop into a structure called the *corpus luteum,* which is essentially a temporary gland. For the first two to three months of the pregnancy, cells of the corpus luteum manufacture the hormones estrogen and progesterone. Estrogen molecules then travel to the uterus, where muscle cells begin proliferating, increasing the size of the uterus so it will be powerful enough to expel the fetus at birth. Until its concentration declines late in pregnancy, the progesterone, in an exquisite example of the body's control mechanisms, inhibits the uterine muscle from actually contracting. As the placenta grows, some of its cells also take on the jobs of making estrogen and progesterone, eventually producing far more than the corpus luteum.

Thus it is the tiny blastocyst—a speck barely visible to the naked eye—that commandeers the woman's entire body, turning it into a life-support system upon which the embryo can depend during development. (The most common test for pregnancy, incidentally, is to look for CG in the woman's blood or urine.)

Only now—about nine days after fertilization but still nearly a week before the next menstrual period is missed—is the true embryo, the humbly named inner cell mass, ready to begin developing. Even now, however, it will be yet another week before a few cells of the inner cell mass reorganize themselves into the first actual structure of the new embryo—the "primitive streak." Still little more than an amorphous lump of about twenty undifferentiated cells, this "inner cell mass" now takes on the thirty-six- to thirty-seven-week task of constructing or, to adapt the old word, conceiving a human infant.

It is a visage that imprints itself on the memory: the face of a human embryo, smooth, pale, almost featureless yet clearly of the human kind. The lidless eyes stare blankly. The tiny hands and arms float in front, motionless, like the stiffly bent arms of a space-suited astronaut drifting weightless against a black sky, tethered by an umbilical cord.

Though relatively few people have seen an actual human embryo, the image is well known in fiction and fact. There was, for example, the special effects space child of the movie *2001: A Space Odyssey,* an enigmatic symbol seen through the gauzy veil of an amniotic membrane. And there are the marvelous photographs of real human embryos by Lennart Nilsson that the original *Life* magazine published nearly a generation ago, not to mention another Nilsson series published in recent years. At just four weeks of development, the photos show us little more than a curled, lumpy form—the head end just discernible, a long tail startlingly prominent. Then after eight weeks, the tail gone, we see the incredible detail of what looks like a little human body. Each tiny finger separate. Eyes with lids. Ears beginning to flare. A little button of a nose. And this whole embryo is at eight weeks barely one inch long, smaller than the last joint of your little finger.

What none of the popular images reveal, however, is the drama within the embryo. Many of the most profound phenomena of human development ment are long finished before there is anything remotely human to see. Just two weeks after the blastocyst has burrowed into the uterine lining, for example, a heart has been constructed and is beating. The entire embryo is no bigger than a sesame seed. Still earlier, just one week after implantation (roughly two weeks after fertilization), the cells of the inner cell mass emerged from their imposed dormancy and began crawling furiously, slithering over one another, pushing between pairs of neighbor cells, tunneling deep into masses of other cells. Like the members of some marvelous acrobatic troupe, each with a precisely choreographed role, the founding cells of the true embryo—the part that will become the baby—assemble themselves into complex formations, creating new architectures that grow in complex elaboration by the day.

The most astonishing acts of human embryonic development happen long before there is anything resembling a face, when the embryo is far simpler than an earthworm. Understand how these early events happen, say cell biologists and developmental biologists, and you will know how all the rest comes to be. For it looks as if much, if not all, of the anatomy of the human body is the result of the same kinds of events happening over and over again but in different patterns and combinations. The biggest changes are in the different suites of genes that are switched on at each step or repressed into dormancy.

As the embryo's cells divide, the daughters go on to assume forms and functions the mother cell could not express. In some cases, one daughter cell pursues one career while its sister cell follows a different path. Often the cells assemble into tissues that serve for only a few days or a few hours and then they die out—victims of an absolutely essential process called programmed cell death, or, in the word that some cell biologists use, apoptosis (from the Greek apo, meaning "away from" and ptosis, meaning "falling"; the second p is silent).

Programmed cell death (PCD) has also long been known to help sculpt the physical form of organisms. The tadpole's tail, for example, shrinks away in a massive wave of apoptosis. Something similar happens in human development. During the fifth week of gestation, a human embryo's hands are flat paddles with no separation between the parts that will be fingers. Then, during the sixth week, four strips of cells in each paddle undergo PCD. Their mission as scaffolding in the construction of fingers completed, they obligingly bow out.

The precision with which embryonic development unfolds has long puzzled developmental biologists. Is the fate of every new cell that arises determined strictly by its genes? Or does the environment around each cell play some role? Sydney Brenner, a pioneering British biologist who first proved the existence of messenger RNA and who did much early research in embryonic development, once compared the two systems to human cultures and dubbed them the "American plan" and the "European plan." Research on the development of various species has shown that both systems operate in nature and each species uses a combination of the two.

For example, the tiny, nematode worm called *Caenorhabditis elegans* was once thought to follow the European plan almost exclusively, each of its cells inheriting a specific function from its ancestors. In the case of embryos, however, the ancestors are not previous generations of whole organisms but the cells within an organism whose division gives rise to a new pair of descendent cells. The impression of an invariant sequence of fates grew out of a remarkable body of research on *C. elegans,* a relatively simple beast, a fraction of an inch long, and one with a very rigid plan of embryonic development. The male's body comprises exactly 959 cells and the body of the hermaphrodite, the only other sex, is made up of 1,031 cells. Though small, a nematode does have various internal organs, a mouth, a gut, gonads, and so on. Because of its simplicity, short life cycle, and ease of rearing, biologists have been able to study nematode development in such detail that they have followed (by watching through a microscope) its development from the one-celled stage to the adult, tracing the careers of each individual cell at every stage of embryonic development. They have worked out a complete "family tree" of the cells, called a fate map. And because all normal worms of the same species followed exactly the same pattern of development, researchers thought *C. elegans* was an extreme example of the European plan. The thing that changed at each round of cell division was not the genes. It is the rule in biology that all genes remain present in all cells of an individual. Instead the combination of switched-on and switched-off genes was passed on. Each daughter cell somehow remembered which genes were functional in its mother cell and would either continue as the same type of cell or add a modification to the inherited pattern, giving it a new form or function.

*Very rigid*

Humans with their trillions of cells of some 200 kinds, by contrast, were thought to follow the American plan as, presumably, do all vertebrates. The fate of each vertebrate cell was thought to depend partly on its inherited pattern of gene activation but also on its circumstances. With each new generation of cells that form in a vertebrate embryo, the next step in differentiation depends on both the inherited pattern of genes that have been switched on or off and the molecular signals reaching the cell from its micro-environment, including its cellular neighbors. The signals that arrive at a given cell depend on its location within the embryo. At least one French scientist calls this the "Swiss plan." Your neighbors tell you how to behave.

Developmental biologists now know that in both worm and human, the process of development is very much the same—a combination of European and American plans. At a succession of points during development, more and more genes become permanently switched off, and there is no going back. At the same times, other genes kick into action, sometimes never to be switched off. The phenomenon of daughter cells acquiring their mother cell's pattern of gene activation is called cell memory. ✓ How it works is not understood. Transplant some pancreas cells into the brain, for example, and they "remember" that they are pancreas cells and keep churning out pancreatic hormones. Take them out of the body entirely, however, and put them in a flask of culture medium without any of their usual chemical signals from other parts of the body, and they forget. Under these conditions, some differentiated cells will relinquish some of their specializations. Many revert to the generalized blob shape of the amoeba and begin crawling around. Yet, as experiments have shown, these cells remember their old specialized role. If the right environmental signals are restored, the form and behavior of the cells shift back toward the specialized role. Some of a specialized cell's properties are retained via cell memory (European plan) and others are only transiently maintained by their circumstances at the moment (American plan).

Mina Bissell of the Lawrence Berkeley Laboratory at the University of California has performed experiments that demonstrate this dramatically. Her work gives a good idea of how the undifferentiated cells can be made to form new shapes and express new behaviors.

Bissell removed a few cells from a female mouse's mammary gland and grew them in a plastic dish of nutrient broth. The cells lost their specialized shapes and became blobs. But, when Bissell switched the cells from plastic to a surface that mimicked the environment a cell would experience within the mammary gland, the cells changed radically. She coated the bottom of their dish with a layer of proteins similar to those that many cells of the body secrete into the spaces outside themselves. These are usually fibrous proteins that form a microscopic, feltlike mat called *extracellular matrix.*

Instead of slithering about as lost individuals, Bissell's mouse mammary cells gathered into clumps. Then the clumped cells crawled over one another until they formed themselves into hollow balls of cells. Then the cells changed their internal architecture so that the inside face of each cell became specialized for secretion. When Bissell added a drop of prolactin, the hormone that stimulates mammary glands to make and secrete milk, the little balls began swelling with mouse milk. The tiny milk glands did not, however, form anything like the ducts that would carry the milk to a nipple.

Experiments of this sort reveal something else fundamental to the process of embryonic differentiation—even cells that have given up their specialized form still carry specialized cell memories. When prolactin is added to cultured cells from nonmammary tissues, they do not make milk. This demonstrates that once cells have become specialized, they are committed to certain forms of activity.

Bissell's experiment shows that a relatively modest change in the cell's external environment can lead to major changes in the cell's behavior. Loner cells had cooperated to transform themselves into something resembling their specialized role in a mammary gland. Somehow the extracellular matrix had sent a message to the cells that activated certain dormant genes. In other laboratories, similar experiments have led cultured cells to organize themselves into primitive versions of other organs, sometimes called organoids. For example, liver cells in a culture flask will assemble into a rudimentary liver, making large, ball-shaped structures that carry out many of the chemical processes typical of the liver.

This task is made easier because specialized cells wear the biological equivalent of name tags—special molecules on their outer surfaces that are worn only by the other cells with which they are supposed to link up. These molecules bind only to other molecules of the same kind and stay bound, essentially welding like-minded cells into tissues.

Also, as has been mentioned, cultured nerve cells will send out long processes that form synapses, connections with neighboring cells intended to relay the electrochemical messages of a nervous system. The cells that normally secrete cartilage, the rubbery gristle that supports the nose and ears as well as other bodily structures, will produce sheets and lumps of cartilage in a culture dish. The cells that give rise to bone will crawl over the bottom of a dish, laying down a material that soon turns into bone. In all cases, the genetic endowment of the cells remained the same, but the environment somehow induced appropriate sets of genes to express themselves.

In the developing human embryo much the same phenomenon, including the crawling of individual cells from one location to another, happens repeatedly. As the cells change position, they alter one another's environments in ways that induce shifting combinations of genes to express themselves. In response the affected cells may multiply in place, migrate elsewhere, or even travel in unison with a sheet or train of neighbor cells. Beginning in the first week after fertilization, the cells of a human conceptus (technically, it is not an embryo until the second week after fertilization) begin the most elaborate of choreographed dances. They glide, turn, and even change costume. As individuals or in groups they move this way and that, climbing over neighboring cells, diving below others. Their movements are dictated by a delicate interaction between their genes—stamped with an evolutionary legacy formed over billions of years—and the evanescent molecular signals of each cell's environment, which is changing by the moment. The genes lay out the possibilities for each cell. The chemical environment selects among the possibilities, and the cell, like a little computer, stores the choice in its memory. The new entry in the memory bank is like a new line of code in the computer's program, affecting everything the cell does in the future. And the new line of code is passed on when the cell stops to divide. Achieving a complete understanding of the billions of interactions among cells in the developmental dance will remain one of biology's greatest intellectual challenges for many years.

About the time a pregnant woman first misses a menstrual period—about Day 15 after fertilization—the cells of the inner cell mass have organized themselves into a different form. By moving into different relationships with one another, they have created essentially two small, hollow balls of cells inside the larger ball. The two tiny inner balls are pressed against each other and also suspended within the larger hollow space by a stalk, the primitive umbilical cord, which anchors them to the trophoblast. The true embryo, even at this stage, is nothing more than the two layers of cells formed where the two balls touch, like two balloons pressed together. This nearly circular, two-layered area is called the embryonic disc. From these two sheets of cells the baby will develop. One ball is the yolk sac, an evolutionary relic of the mass of cells loaded with fat and protein that provide nourishment for embryos growing inside eggs. It will soon be used up, and the ball will essentially shrivel up into the belly of the embryo. The other ball is an open space that will become the amniotic sac. Embryologists traditionally mark the beginning of the formation of the embryo—they call it embryogenesis—from the point at which a kind of groove, called the primitive streak, forms in a straight line over the surface of the embryonic disc. This defines the future head-to-tail axis of the body and is the first sign of the formation of the gut. The nonembryo parts of the overall structure will eventually develop into such organs as the amniotic sac, the umbilical cord, and the placenta, growing into nearly as large a mass of living tissue at birth as the baby.

All these cells, whether destined to be loved as a human being or discarded as afterbirth, contain identical suites of all human genes—identical inheritances from a mother and a father. However extreme the difference in their fates, the cells that develop from a fertilized human egg are all actors in a single, tightly coordinated process of human development—surely one of the most complex phenomena in the known universe.

Just one of the organs that will be created, the brain, is often called the most complex structure in the universe. How much more complex, then, is the process that constructs not only the brain and nervous system but the immune system, the reproductive system, the hormonal system, the muscles, the kidneys, the liver, and dozens of other organs, all of which interact with one another and with the brain to make a complete human being? Remarkably, most of the anatomy of human embryonic development has been well known for decades, much of it learned from painstaking dissection in the nineteenth century of dead human embryos. The mechanics can be described in considerable detail. What remains a mystery in most cases is the cause of each event at the molecular and cellular levels. What is the mechanism of cell memory? What is the signal that tells a given cell or group of cells to crawl to some other part of the embryo? What cells issue the message that tells other cells to divide or to hold off dividing? Such are the questions at the frontier of the field that is today called molecular embryology. The challenge that lies before the field is perhaps the greatest in all of the life sciences—to learn how one fertilized egg cell, containing a microscopic dose of molecules of various configurations, can organize itself and its progeny to take up raw materials from its environment (the mother's womb in the first nine months) to

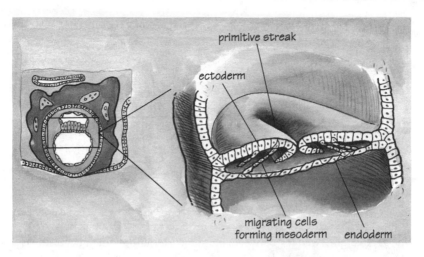

**Figure 8.1** Within the embryonic disc, the first structural change to create an organism is formation of the primitive streak. Cells of one layer crawl under their neighbors and form a middle layer, the mesoderm, between the two original layers, ectoderm and endoderm. This activity begins at one end of the disc (forming the primitive pit) and proceeds in a straight line toward the other end, forming the streak. This axis will define the structural precursor to the embryo's spinal cord and backbone.

construct the exquisitely detailed architecture of the human body. Somehow the message of the genes—carried in the linear, one-dimensional code of DNA—is translated into a three-dimensional organism.

After the formation of the embryonic disc, the first step in the development of what will become the baby is the formation of what developmental biologists call the primitive streak (Figure 8.1). This is also the step that confers a degree of individuality to the embryo. Although genetic individuality was acquired at the zygote stage, that step did not establish whether the newly formed set of genes would result in just one person or more than one. Morulas sometimes break into two or more parts, each developing into a separate pre-embryo, each with its own placenta. Or the separation can happen a few days later, to the inner cell mass, and also become two pre-embryos, but they may then share one placenta. In other words, events after fertilization will decide whether the initial genetic individual gives rise to one baby, to identical twins, or to still more genetic duplicates.

Cells from opposite sides of the disc begin migrating toward an imaginary line running down the center of the disc. Like waves coming together from opposite directions, the cells become heaped up in a ridge along either side of a groove—the primitive streak. Actually the waves meet first at one side of the disc, at the edge nearest the stalk connecting the two balls to the rest of the conceptus. Between the two ridges is a pit, but as more of the opposing waves of moving cells meet, the pit lengthens into a groove. This is the structure called the primitive streak.

As the ridges meet, a few cells from each side crawl into the groove, diving into a slight opening between the ridges. The motile cells then crawl back under the ridges, creating and occupying a space between the ectoderm—the layer from which they came—and the layer below, called the endoderm.

The place where the ridges first meet will be the tail end of the embryo. Over the course of a day or so, the streak lengthens toward the middle of the disc, toward what will be the head end of the embryo. More cells of the upper layer crawl between the two layers, essentially plowing a groove toward the middle of the disk. A moving sheet of ectoderm cells dives into the groove and wedges between the two original layers. The result is a third layer of cells called the mesoderm.

Cells derived from these three fundamental layers will be the founders of all the major organ systems, as different patches of cells, behaving like those that formed the mesoderm, sever their junctions to their old neighbors and slither off to take up new positions and to change their suites of active genes. Ectoderm cells will, for example, become the epidermis, the skin's outer layer, and the entire nervous system. Mesoderm cells will give rise to the inner layer of skin called the dermis and such structures as muscles and bones. Endoderm cells will develop into the gastrointestinal system, the liver, and the lungs, among other organs.

Embryogenesis is a process that grows in complexity by the hour, at least during the early weeks. Just one day witnesses such dramatic transformations that experts can, for example, tell a seventeen-day-old embryo from an eighteen-day-old one. And a week spans epochal changes. At various points, certain groups of cells get up and crawl off to different destinations. Upon arrival, they settle down and proliferate. Sometimes whole sheets of cells slide around in unison to place new layers in other locations.

Each new layer or fold creates a new environmental context in which shifting patterns of genes are expressed. As each new suite of genes comes into play, the environment for neighboring cells changes, inducing them to express still other patterns of genes. As a result, groups of cells become specialized in many ways.

Certain mesodermal cells, for example, curl themselves into a network of tubes that become the primitive arteries and veins. The circulatory system does not grow and spread out like branches on a single tree. Instead, many small islands of blood vessels appear in scattered sites around the embryo. Each tube lengthens and sprouts buds that branch. Then the blood islands, as they are called, meet and fuse to form one large network. The blood islands arise not just in the embryo but throughout the primitive placenta as well. By Day 22, the mesodermal cells, wrapped around one of the larger tubes, change into muscle cells, full of long bundles of actin filaments, the same cytoskeletal element used when individual cells crawl. Next to the actins are parallel bundles of myosin, filaments with motor molecules at the ends. The myosins begin pulling on the actins, making the long cells contract. The muscle cells begin contracting rhythmically and together form the primitive human heart, a one-chambered pump.

While the forerunners of arteries and veins were forming, they were also extending themselves into the yolk sac to tap the food supply there. At the same time blood cells were also forming. Cells trapped inside the blood islands transformed themselves into blood cells, as did those in the primitive umbilical cord. In later life, blood-forming cells live within the bone marrow.

Now, just one week after a woman notices her first missed menstrual period, a simple heart is pushing embryonic blood cells through an embryo the size of a sesame seed, out the unbilical cord, into a surrounding placenta the size of a pea, and back to the embryo.

Step by step, the plan of embryonic development continues, creating an ever more complex structure of interacting cells, each reading out portions of its genetic code according to its "cell memory" and whatever molecular signals reach the cell to convey messages to the DNA.

Further evidence that cell-to-cell signaling must be happening emerged in 1990 from the new findings that at last begin to answer a question raised by a classic embryological experiment done in the 1920s. Back then two German embryologists, Hans Spemann and Hilde Mangold, were tinkering with animal embryos at the blastula stage, the time common in all animal development when the conceptus is just a hollow ball of cells. Spemann and Mangold found that if they transplanted a certain group of cells from one blastula to another, the recipient blastula would produce two complete embryos. It did not happen when any old bunch of cells was transplanted, only cells from a key part of the blastula. The result was the original embryo plus an extra one centered on the transplanted cluster of cells. The two researchers concluded that the transplanted cells were special, functioning as some kind of "organizer," capable of sending signals that induce otherwise idle cells of the blastula to differentiate in a coordinated pattern.

Jim Smith, a molecular embryologist at the British Medical Research Council's National Institute of Medical Research in London, discovered a substance that may be a signal from the organizer. Working with frog embryos, Smith isolated a protein, which he calls activin, that has profound effects on other frog cells. When he exposed undifferentiated frog cells in a dish to a low dose of activin, they changed into skin cells, one of the ectodermally derived cells. A somewhat higher concentration of activin turned undifferentiated cells into muscle cells, which normally stem from mesoderm. And the highest doses of activin turned the cells into mesoderm that then proved able to induce ectoderm cells to roll themselves into a neural tube, the founding structure of what develops into a brain and spinal cord (Figure 8.2).

A number of other labs have come up with comparable findings that promise to open new windows on embryonic development that may reveal the signals that launch relatively generalized cells onto the developmental programs that can create a fully formed organism.

One of the most profound observations of embryology—one that hints at ancient evolutionary links among all life forms—can be made at about this stage, four weeks after fertilization. At this point, where there is a clearly defined head end and the beginning of buds that will be the arms and legs, the human embryo is almost indistinguishable from the embryos of most other vertebrates.

As embryologists have remarked for more than a century, an early human embryo looks very much like an embryonic pig or lizard or bird or any of a host of other animals. At this stage, for example, human embryos, like the others, have gill pouches, structures that in fish become gills, and a tail. Eventually the human tail shrinks away and the gill rudiments develop not into gills but into certain bones, muscles, and other structures of the human face, ears, and neck (Figure 8.3).

Biologists believe that these structures arise because human beings retain the ancient genetic information—shared with the fishes—to construct their rudiments. But as development proceeds, other sets of genes come into play. In fish, the environmental signals received by cells in the

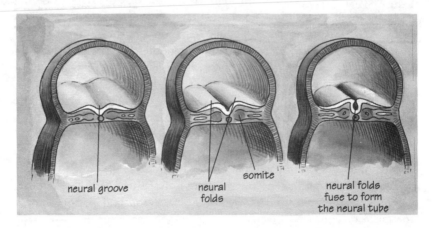

neural groove     neural folds     somite     neural folds fuse to form the neural tube

**Figure 8.2** During the third week after fertilization, layers of the embryonic disc curl and fold to create parallel tubes along the embryo's axis. Here the formation of the neural tube, destined to become the brain and spinal cord, is suggested.

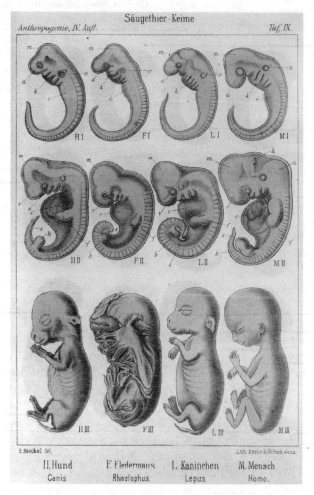

**Figure 8.3** In 1874 the great German biologist Ernst Haeckel published this comparison of embryos from four species. From the left they are dog, bat, rabbit, and human. (From E. Haeckel, *Anthropogenie, oder Entwickelungs-geschichte des Menschen.* [Leipzig: Engelmann, 1874] Courtesy of the Bodleian Library, Oxford)

gill arches trigger genes that dictate the development of gills. In humans those genes are either absent or inactive or they have mutated into something different. The cells in human gill arches carry genes that when activated by signals in that region of the embryo cause the tissues to remodel themselves into something entirely different from gills.

Embryologists say it is hard to learn how all these complex events happen, but experiments like Smith's on activin in frog embryos and

Bissell's on the mouse mammary cells that formed rudimentary milk glands give important clues. They show that a cell's environment is a powerful controller of gene expression.

One type of cell signal that biologists have long thought to play a key role conveys information about the cell's position relative to its neighbors. Studies of fruit fly embryos, for example, show that cells at the head end of the embryonic disc secrete certain chemicals that diffuse toward the tail end, setting up a gradient in the concentration—higher near the head, lower near the tail. Other chemicals diffuse from tail to head and still others from top to bottom. Although researchers suspect that a similar mechanism operates in vertebrate embryos, no evidence of it has turned up despite much searching. The vertebrate, and therefore human, method of specifying a cell's position in the overall embryo remains a mystery.

In the fruit fly, each cell in the early embryo receives outside information in the form of specific molecules from several directions. The chemicals may work like a system of map coordinates, telling each cell its position in the embryo.

This interplay between genes and environment leads to a profound conclusion. Heredity, contrary to what has sometimes been asserted, is only partly destiny. Environment is also destiny. This is why, for example, so-called identical twins—people who share identical sets of genes—are not really identical. Yes, they often look very much alike and they often live uncannily similar lives. The impression that extreme similarity is inevitable is created by scientific studies or journalistic accounts that focused on those twins who seemed most identical. The fact is that there are plenty of identical twins who have differences, ranging from the subtle to the radical. For example, one twin may be big and healthy while the other is small and sickly. One common explanation is that the two embryos shared one placenta but that one twin claimed a bigger share of the blood supply. More subtle differences usually show up in mental abilities and personalities. Here the environmental difference can occur after birth in such things as the richness of a baby's surroundings, but it can also happen before birth when individual cells of the brain are being guided to sprout more, or fewer, tentacles to make connections with other brain cells. Neither do genes encode the exact pattern of wiring that links brain cells.

"The 'totality' of a person is not genetically programmed into a newly fertilized egg cell," says Charles A. Gardner, a developmental biologist at the University of Michigan medical school who studies genetic influences on the development of the brain. Gardner likes to remind anti-abortion advocates of that fact whenever they say that a fertilized egg is a full-fledged person because it represents the "totality" of everything that a person will be. "People are just too darned complicated for the pattern of an individual fingerprint or, more inportantly, of an organ as complex as the human brain to be specified in that cell. Biology leaves some of the details up to the whims of embryonic development. Rewind the tape of development, play it over again and you would end up with a different person, with different connections in the brain, with different fingerprints."

For an embryo to develop successfully, however, each specialized cell type must arise only in the right place. If the cells that should form the heart started to specialize as eye cells instead, the embryo would soon die for lack of blood circulation. Many kinds of abnormalities can arise in development but this kind of misplaced cell type is rare. Or if those primitive heart cells produced lots of actin but little or no myosin, they couldn't contract. For development to proceed, groups of genes must begin to function at the same time in cells in the right place. Other genes must function at different times but in the correct sequence and only for the proper duration. All of this means that life must have some mechanism for coordinating the expression of suites of genes, all of whose products are needed to carry out a given biological function. An extraordinary clue as to how this coordination might happen has come from research on that old standby of geneticists early in the twentieth century, the fruit fly.

The beauty of the fruit fly is that it is small, lives happily in bottles, and breeds new generations so quickly that biologists can easily find new mutations arising in a population and track down the very gene that changed. So many flies were raised in laboratories and examined so closely that, over the years, fruit fly researchers reported some of the most astounding kinds of mutations ever witnessed in any species. One researcher, for example, spotted a fly with an extra pair of wings. This mutant fruit fly, which normally would have one pair of wings growing from its thorax, had the equivalent of two thoraxes between its head and its abdomen, each with a set of wings. Even more bizarre, somebody else found flies with legs growing out of their heads in place of antennae (Figure 8.4). In nonscientific ages, freaks such as these were dismissed as uninteresting or as the work of a devil or some other supernatural power. But in a time when even the mystery of embryonic development is seen as the result of natural phenomena, and hence is potentially understandable, fruit fly geneticists took these oddities as welcome experiments of nature. They were seized upon as windows on a long-hidden process.

Despite the fact that very different regions of fly anatomy were involved in these mutations, they had in common the fact that the abnormalities consisted not of badly malformed parts of the body but the development of a perfectly normal body part in the wrong place. If embryonic development is the result of each cell getting chemical messages as to where it is in the body and then expressing the appropriate genes, how in the world could a cell near the eye get signals telling it and its cellular progeny to express the genes needed to become a leg?

**Figure 8.4** On the right is a normal fruit fly, its hairlike antennae mounted just inboard of the two red eyes. On the left is a fly with a mutation in a gene that, essentially, made the cells that should give rise to antennae behave as if they were farther back, where the legs should be. As a result, the suite of genes that guide development of a leg were switched on instead of antenna-making genes. (Photo courtesy of Professor Matthew Scott)

The beginning of an answer has emerged from the discovery of the genes that mutated in the freak fruit flies. It turned out that in every case, very few genes were altered in each fly, often only one. But the crippled genes were always ones that contained one particular type of genetic sequence—a series of 183 bases embedded within the thousands of bases that make up the whole gene. This means that the proteins dictated by those genes all contain similar regions of the same sixty-one-amino-acid sequence. Molecular biologists have found that the shape of the part of the protein dictated by those bases is able to bind to DNA. This implies that all the different proteins that contain this sequence are proteins that do, in fact, bind to DNA, presumably to regulate gene expression.

Molecular biologists have concluded that these proteins work as gangs of identical molecules, each binding to one of several functionally related genes, either blocking or facilitating the expression of all of them simultaneously. The mutation that made a leg grow where an antenna should be was in a master gene that made a protein which normally binds to

DNA in several places, blocking the expression of genes that were supposed to remain dormant. But when that master gene is defective, its protein cannot work properly and, in effect, the brakes are released from all the genes it is supposed to hold in check. Without being repressed, all the now-unbound genes—which all plays roles in the development of one particular structure—are free to follow the "European plan," carrying out a coordinated genetic program to construct some set of tissues, say a leg, even if the cell where the mutation arose happens to be in the animal's eye.

In other cases, researches have found genes containing similar sequences that do not make repressor proteins but stimulator proteins. Either way, the concept is the same. Since a given protein might bind only to a particular set of genes—repressing them all at the same time, or activating them all—the proteins provide a mechanism for coordinating the activity of a set of genes that, when expressed together, cause the affected cells to develop some major structure of the organism.

The discovery of genes containing the similar sequences was at first a curiosity. Perhaps fittingly, such genes were given a curious and unfortunately obscure name: "homeobox" genes, from the Greek *homeo* meaning "similar" and "box," a molecular biologist's colloquialism for a string of identical or nearly identical bases found within several genes. Originally homeobox genes were thought relevant only to insect development. There seemed to be a definite relationship between the way they operated and their position in a particular one of the many segments that make up an insect's body. Cells presumably receive information about which segment they inhabit from the gradients of signal molecules diffusing in various directions from one part of the embryo to another. One possibility—only speculation at this point—is that if the cells in a particular segment get signals appropriate to a segment that is supposed to make legs, those signals somehow switch on a certain gene containing a homeobox, it activates the master leg-making genes, and, in effect, the software for making a leg is released to play itself out. In other words, the event that happens normally in the right place has the same effect as the mutation that arises in the wrong place. Some of the more bizarre mutants seemed to be the result of defective homeobox genes "misinterpreting" which segment they were in. For example, the flies with two sets of wings were simply ones in which two adjacent segments, instead of just one, behaved as if each thought it was the only one in the right place to sprout wings.

But it was soon discovered that insects were not the only organisms to have such genes. Once molecular biologists learned how to look for them, they have found homeoboxes in all higher species tested, including human beings. Scientists have been stunned to learn also that the homeoboxes are all very similar in genetic sequence from one species to the next. In fact a mouse homeobox is more like the homeobox in the equivalent human gene than the mouse hemoglobin gene is like the human hemoglobin gene. The similarity among homeobox genes means that their resulting proteins all contain one module that is virtually the same in all proteins and, therefore, confers a common functional role on all the proteins.

The great similarity among homeoboxes is striking because evolution is constantly making tiny, random alterations in almost all genes and the longer two species have been separated from their common ancestor, the more differences they have in otherwise comparable genes. In other words, mouse hemoglobin and human hemoglobin, though somewhat different, are more alike than are human hemoglobin and say, frog hemoglobin. Biologists expect that the only mutations that survive the ruthless weeding-out process of natural selection are those that change the gene into one that is more useful to the organism because it confers some advantage in survival or, at least, that the mutation imposes no great disadvantage. The extraordinary stability of the homeobox—it is little changed over hundreds of millions of years of evolution—means that it must play a crucial and indispensable role and that most deviations from its proper sequence are harmful.

Human embryonic development appears from these findings to be very intimately dependent on genes containing a homeobox sequence. In yet another demonstration of the fundamental similarity among all living things, the universality of the homeobox shows that once evolution came up with a way of coordinating families of genes so that they act in concert, it kept that method for every new species that arose. The genes themselves may have different consequences from one species to the next (actually the differences are usually minor), but they appear to be controlled in exactly the same ways.

Early in human development the embryo takes on an appearance that is reminiscent of the segmented pattern of insects and indeed of many evolutionarily early organisms such as earthworms. In the third and fourth weeks, along the length of its head-to-tail axis pairs of bulges appear, one on each side of the midline. The bulges, which look like parallel strands of pearls, are called somites, from Greek *soma,* for "body." Eventually forty-two to forty-four pairs of somites form. Early in embryonic development the somites play key roles, each being a domain in which certain developmental events occur. Each somite becomes the site of the formation of a spinal vertebra, certain muscles, and, in some cases, ribs. Also a patch of skin develops from each somite. The chief vestige of this early segmentation in adults is the spinal column and the ribs. Curiously, however, each vertebra in the spine represents not one somite but two half-somites. In other words, the center of each somite becomes the cartilaginous disc between the vertebrae and the ends of the somite on either side give rise to the half of a vertebrae on either side. Thus each vertebra is the product of two adjacent somites.

While no one has yet linked somites to the action of homeobox genes, there is every reason to believe that human homeobox genes play a "master gene" role in coordinating embryonic development as they do in fruit flies. Unfortunately, says Joel M. Schindler, a biologist formerly with the National Institute of Child Health and Human Development, "we remain woefully ignorant of how these genes accomplish their biological roles." Schindler maintains that the research that is still exploring fruit fly homeobox genes "may have direct relevance for improving the human condition," in which, for example, more than 250,000 babies born in the United States each year are afflicted with congenital malformations, a high proportion of them the the result of flaws in the genes that guide embryonic development.

"You think of all these things that have to happen just right. You think of all the steps that could go wrong. The surprising thing is not that it sometimes goes wrong but that it ever goes right," says Richard Tasca, an embryologist at the same institute, which is one of the National Institutes of Health.

Still, the rare times when development does go awry can be heartbreaking. The sad catalog of hundreds of different kinds of birth defects—from the surgically repairable harelip to the invariably fatal failure of the brain to form—is gradually being understood as the result of problems that arise at specific stages when embryos pass through crucial steps in development.

For example, Thalidomide, the drug that left babies with "flippers" for arms, is now known to have worked its effect before the embryo was forty-eight days old because the arms and hands of a normal embryo attain nearly normal proportions, though of course not their final size, by then.

Even more limited defects, such as two or more fingers fused together to produce a deformation called "lobster claw," can be traced to problems during Days 42 and 43, when the embryo is about 7/8 inch long. This is the time when webbing between embryonic fingers is supposed to disappear. The cells of the webbing are programmed to die at this point and disintegrate, leaving a space between the fingers, but in rare cases they don't. How those cells "know" they are to perish for the sake of the other cells has been one of life's great mysteries, but current research is leading to an understanding—the phenomenon cited earlier of programmed cell death. In a nutshell, all cells are genetically programmed to commit suicide and are prevented only by receiving "social controls" from their cellular neighbors—chemical signals emitted by cells to keep others in their community from self-destructing. Some scientists call these chemicals "savior signals." If those signals stop—as they do not only during development but at many times in a lifespan and at various places in the body—the cell activates its genetically determined ritual suicide. The cell shrivels up, breaks into pieces, and is eaten by its neighbors. The process, called apoptosis, is very different from pathological cell death, or necrosis. A cell that dies because of illness or injury usually swells up and bursts, spilling its contents, including powerful digestive enzymes that are normally sequestered inside the cell's digestive organs. Necrosis often triggers an inflammatory reaction that can kill neighboring cells and spread. It is not good for the body. Apoptosis, on the other hand, is neat and clean. It never harms other cells.

"I think the evidence has become quite strong over the last few years that this happens quite commonly," says Martin C. Raff of University College, London, a leader in the study of programmed cell death. "The natural fate of animal cells is to die. This is part of the social contract that makes it possible for a large number of individual cells to live together harmoniously as a single organism. Some of them have to make sacrifices."

 Early in development, sometimes less than a week after fertilization, one of life's most dramatic surprises can arise. The inner cell mass may separate into two inner cell masses. A single "conception," as we saw earlier, is now destined to become two human lives—identical twins. Sometimes the two cell masses remain joined in some way and the result is not only surprising but often tragic—"siamese," or conjoined, twins. They may be essentially two complete individuals sharing a nonvital region of flesh or they may be as close as two heads on one body.

Twinning is an experiment of nature. Embryologists have done the opposite experiment on animals, fusing two or more mouse embryos into what will develop as one mouse. For example, when an early conceptus of a white mouse is pressed together with a comparable conceptus of a black mouse—typically each is at the eight- or sixteen-cell stage—and the combination implanted in the uterus of a female, she eventually gives birth to a two-toned mouse that had four biological parents. "It's things like this that challenge your idea of what an individual is," Tasca says.

Embryo fusion experiments have shed much light on the mysteries of development. The patchy black and white mice, for example, reveal that contiguous regions of the body are made up of cells that descended from only a few embryonic precursors. The cells mingled and, though maintaining their different identities, cooperated to construct one perfectly formed organism. If a black cell happened to be near the right shoulder, its cellular descendants that became skin cells sprouted black fur. If the neighboring embryonic cell was white, there would be an adjoining patch of white fur

Embryo fusion experiments are another way of showing that although cells can express only the genes they have inherited (making the difference, for example, between black and white fur), the factors that determine which genes are expressed (making the difference between a fur cell and, say, a muscle cell) lie outside the individual cell.

The nervous system is the first major part of the human body to arise in embryonic development. The process begins just two and a half weeks after fertilization. Thus the foundation for the human brain, often exalted as the crowning achievement of evolution, is laid early. But, of course, so is the foundation for the mouse brain, the bird brain, the lizard brain. What makes the human brain special is that its development continues for so long—all through gestation and even beyond birth. In fact, the human brain does not complete its anatomical development, achieving its full complement of cells and the wiring that finally links all its parts, until several years into childhood.

On about Day 18, when the embryo is barely ¹/₁₆ inch long, the neural groove begins to form—the first step in constructing the spinal cord and the brain. The process follows the axis determined by the primitive streak and it involves sheets of ectodermal cells shifting and bending to make two parallel ridges and a groove between them. The ridges get higher and higher until they curl toward one another and touch to form a tube. Then the crests of the ridges curl toward one another until they meet. As the ridge crests touch, they close the hollow space of the groove below them into a tube, forming the embryonic spinal cord and brain.

Closure of this so-called neural tube begins in the middle of the embryo's axis and proceeds in both directions, as sheets of cells curl themselves into a tube, steadily sealing off more of the groove. If this seemingly simple phenomenon goes wrong, it can be the cause of major birth defects, such as spina bifida, in which part of the spinal cord bulges through an opening in an infant's back. The bulge is at a point where the neural tube's curled edges failed to meet. Even more severe—and usually fatal—if neural tube closure does not proceed all the way to the head end, the brain fails to develop properly. In the worst case, a child can be born with anencephaly, the absence of a brain—a fate that is sealed between Days 23 and 26.

Actually, formation of the neural tube is a little more complex. As the tube closes, one of the most extraordinary classes of cells develops along the crests on each side of the neural groove. Just before the crests meet to close the tube, these so-called neural crest cells jump off and crawl away to pursue their own fates in other parts of the embryo. As the neural tube closes, it sinks beneath the surface and the remaining ectoderm, still maintaining a relatively flat surface, closes over it. Between the protective ectoderm and the neural tube lie the scattering armies of neural crest cells.

These cells migrate off to an unusually wide variety of tasks, some to become various parts of the nervous system (including all the sensory nerves), others to help construct the teeth, still others to make some forms of cartilage and the bones of the face, and yet another group to form parts of the adrenal glands. Some neural crest cells even migrate out to what will become the skin, where they take up permanent residence and manufacture the melanin that gives skin and hair its color. This broad array of very different roles has long astounded and perplexed developmental biologists, for when the cells leave the neural tube, they all appear to be identical, unspecialized cells. Yet after they take up their respective duties throughout the organism, each type of specialized cell has its own characteristic suite of active genes. The genes for making melanin, for example, are shut off in neural crest cells that give rise to bone and the bone-making genes are disabled in melanocytes, the melanin-making cells.

Are the cells already committted to their fates before they leave the neural tube or do they get the message along the way or not until they reach their ultimate homes? Nicole M. Le Douarin, an embryologist at the National Center for Scientific Research (CNRS) in Nogent-sur-Marne, France, has removed individual neural crest cells from chick embryos, grown them in culture, and tested them to see what genes they express and what specialized forms they take on. She found that some of the cells, even before they jump off the neural crest, are already committed to certain specialized roles. For example, if removed and cultured under certain conditions, they quickly change shape to look like nerve cells. But other cells in the neural crest have not yet made any commitments. When they begin migrating from the neural crest, they can be induced to express any of the full range of genes appropriate to all of the possible specialized roles. But as these cells make their way through the embryo, certain genes are disabled, one by one. Le Douarin found that neural crest cells that have been migrating for several days have usually lost the use of all their genes except those appropriate for one specialized role.

In the third week, shortly after construction of the primitive nervous W3 system begins, the pace of development speeds up dramatically (Figure 8.5). The heart and blood vessels, as we have seen, begin to form at about the same time. The heart begins beating at the beginning of the fourth W4 week. Strangely, the heart forms in front of the head. Two mushroom-shaped bulges grow out from the brain to begin development of the eyes. Toward the end of the same week the embryo sprouts four buds that will become arms and legs. Also during the fourth week, the embryo that lay flat as the embryonic disc now begins to curl, head to tail, into a tight C shape. Among other things, this maneuver neatly tucks the heart downward into the chest region. At the same time the embryo curls around the yolk sac, incorporating it as the primitive gut. The first days of the fifth W5 week see the mouth begin to form, and pits for the nostrils. The visage, however, is grotesque by the standards of the fully developed. Eyes are on opposite sides of the head. Nostrils are almost as far apart. The mouth is

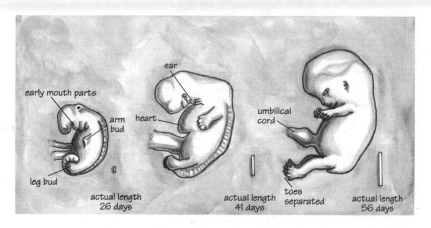

**Figure 8.5** Growth of a human embryo from roughly the end of the first month, when the embryo is the size of a sesame seed, to the end of the second month when it attains the size of a kidney bean. Despite its small size, the embryo at this stage has formed the rudiments of all major organi systems and is traditionally called a fetus from now on.

nearly as wide as the head. By the end of that same week the limb buds have grown longer and developed flattish paddles at the ends. Also in the fifth week the head grows dramatically, largely because the brain has begun enlarging, a process that will not be finished until several years after birth.

The skeleton begins to appear in the fifth week as special cells cluster in dense aggregations to form islands throughout the embryo. The islands are essentially models of the bones to come. Some islands have cells that secrete the tough, fibrous proteins that make cartilage, which is later replaced by bone. In other islands, bone is made directly. In these cases the cells secrete a protein matrix called pre-bone. This matrix provides a surface on which calcium phosphate, which is in the fluid between cells, can quickly crystallize. As more and more calcium phosphate molecules infiltrate the pre-bone matrix, it gradually turns into hard bone. Eventually the cells that secrete pre-bone become surrounded by bone and are trapped. The cells, however, remain alive by maintaining long, thin channels through which they extend tentacles that link one to another and all to the nutrients and oxygen of the outside world. As the embryo grows, so do its bones. Growth is possible because the pre-bone-making cells are periodically released from their prisons by closely related cells that chew their way through existing bone, dissolving the calcium. Such cells constantly tunnel through bone, releasing the trapped cells, which then lay down new pre-bone. The two types of cells work in concert, gradually remodeling bones, making them longer and thicker as the embryo grows and repairing fractures in later life.

At various moments during embryonic development, another curious phenomenon has been happening. Some of the genes contributed by one parent or the other are rendered permanently inoperable. This is highly unusual, for in the vast majority of cases, the cell has two copies of every gene, one from the mother and one from the father. Both are used. But in the case of a few genes, either the maternal one or the paternal one will be chemically disabled by the embryo cell. Cluster of molecules called methyl groups are attached to the gene, rendering it nonfunctional. This inactivation will be passed on with each round of cell division so that, from that point on in the life of the individual, that gene will rarely be available.

Geneticists are just beginning to understand the consequences of this phenonenon, called genomic imprinting. It is believed to play roles in some relatively obscure diseases such as Beckwith–Weidemann syndrome and Prader–Willi syndrome. The story is a little clearer in the better known Huntington's chorea. If the defective gene is inherited from one's father, the disease can begin in adolescence and is more active than if the same gene comes from one's mother, in which case the onset of symptoms may not begin until middle age. The difference is thought to result from differences in the number of methyl groups attached to the gene.

By the end of the eighth week, all major organs and anatomical structures are present, at least in rudimentary form—so clearly so that anyone who sees an eight-week embryo would immediately recognize it as human. It has a torso with arms and legs. Even the fingers and toes are separate and have nails. The face, however, still looks somewhat bizzarre. The eyes, which arose at the sides of the head, are still moving toward the front. (Actually it is more likely that the eyes do not move so much as the rest of the head behind the eyes swells.) The widely separated nostrils have not quite joined to make one nose. The mouth is a slit with no lips. Not until another month or two have passed will the developing face attain "normal" proportions.

Still, at this eighth week—when the mother will have missed her period for only the second time and may not even realize that she is pregnant—when the embryo is about the size of a pecan (about an inch and a quarter long), the foundation of all organs and structures have been laid down. At this point medicine traditionally drops the term "embryo" and starts calling the unborn a "fetus." Virtually all the major migrations of undifferentiated cells to new sites within the embryo have been completed. The cells of the three founding layers (endoderm, mesoderm, and ectoderm) will have arrayed themselves over, under, around, and through one another in a kaleidoscopic variety of relationships. Scores of distinct microenvironments will have been created, each inducing special sets of genes to change the formless cells in those sites into specialized shapes and functions. And, according to one theory of the central role of programmed cell death, cells in each of those microenvironments are emitting "social control" signals that stave off one another's readiness to commit suicide. Through this mechanism each region within the developing body ensures that the cells that survive in it are the ones that are supposed to be there. By the eighth week there will have been established a microenvironment in which each organ, each bone, each muscle will have begun to form.

From this time until birth—thirty weeks later—the most significant changes are growth in size (which happens at different rates in different tissues, thus remolding the proportions of the fetus into those of the infant), and the development of more complex detailed structures within the enlarging organs.

It is during these weeks that the fetus begins to exhibit the first, rudimentary signs of what can be called behavior. The power to move is acquired as the nerves and muscles develop to the point where simple forms of muscular contraction are possible. An eight-week-old fetus moves its arms and legs, however, only in sporadic, aimless ways. This is motion devoid of function. The muscles are gaining the ability to act in unison, so they may flex an arm or turn the head. Embryologists know this because of experiments performed many tears ago on aborted fetuses that were kept in dishes of warm water and observed for a few hours before they died. In this earliest phase of behavior, the fetus is not responding to any stimulus or carrying out any intentional behavior. We can know this with certainty because at this time the fetus's brain is not yet "wired up." There are nerve cells in the brain but they have not yet made the links with one another—called synapses—that are essential for even the most rudimentary mental functioning, including for pain perception.

Shortly after the eighth week of development, however, there is evidence that muscles and nerves become well enough developed to produce reflex muscular contractions, simple responses to outside stimuli. These are movements under the command of a very simple nervous system that still does not include the brain. Reflex behaviors can happen on the basis of as few as two linked nerve cells—a sensory nerve cell to pick up the initial stimulus and relay it to the spinal cord and a motor nerve cell to receive the signal in the spinal cord and carry it to a muscle. Neurologists call this a reflex arc. Reflex responses also occur in adults, such as when the doctor taps the tendon just below the kneecap or when a person jerks an arm back from a hot stove even before the pain sensation arrives at the brain.

Experiments done some fifty years ago on aborted fetuses (they were undamaged by the procedure but doomed because there was no way to keep them alive) showed that by about eight weeks the fetus has simple reflex arcs linking the mouth to the neck muscles. When researchers brushed a hair against the fetus's lips, it slowly turned its head away from the stimulated side. Dissections established the presence of a reflex arc linking lips to neck muscles. (A more developed, and reversed, version of this reflex will help the newborn baby find its mother's nipples. In what is called the rooting reflex, newborn babies automatically turn their mouths toward anything that brushes their cheeks or lips and begin sucking.) In the eight-week fetus's behavior there can be no conscious involvement because the brain cells do not become interconnected for at least another ten weeks.

Although the eight-week fetus's motions are too slight to be felt by the mother at this early stage, this will change. At some point, usually between the seventeenth and twentieth weeks, the fetus's movements become powerful enough that the mother can feel them. Centuries ago this stage was called "the quickening." In those days the attitudes of, for example, Christians toward abortion were generally more tolerant than they are now and abortion was allowed up until the quickening because, according to theologians, that was the moment in which God installed a soul in the fetus. Until that the moment "the quick" was still, in theological terms, "the dead." This ancient belief offers yet another sign that the autonomous motion of an organism was seen as divinely inspired.

Perhaps no question of human development is more relevant to the abortion debate than the matter of whether the unborn has any kind of inner experience, even of the most vague and rudimentary sort, and whether it has any ability to feel pain or even a sense of discomfort. Obviously, this is not an easy thing to find out. But the study of embryos and fetuses does offer some major clues, if one accepts that before there can be anything like feeling or thought or experience, there must be some kind of functioning nervous system. Indeed, the concept of "brain death," now widely accepted, holds that if a person's central nervous system has ceased to function—as evidenced by the failure to transmit signals within the brain—that person is dead. The heart may keep on beating, but without a central nervous system, that person has ceased to exist. In the rather stark vernacular, a human being without brain function is a "vegetable."

Given the general acceptance of those ideas, some scientists and others who grapple with the abortion debate suggest that a similar criterion be applied to the unborn. If there is not yet a functioning central nervous system, they argue, the unborn is incapable of any kind if inner experience, including the sensation of pain, and therefore has not yet become a person in the fullest sense of the word. Until the human embryo or fetus develops the mental capacities that are lost in a brain-dead person, it too would seem to be a vegetable.

Any inner experience that requires a brain, such as feeling pain, is unlikely in an embryo younger than twenty weeks. And it may not even begin to be possible until around twenty-five weeks. It does look as if some form of thinking can happen by around thirty weeks. First of all, synapses do not begin to link the nerve cells in the thinking part of the brain, the wrinkled outer layers called the cortex, until approximately twenty weeks. Because the human brain's extraordinary abilities are the product not of a mass of isolated cells but a system of cells linked in an extraordinarily complex web of interconnections, no thought or perception can arise until they are interconnected. The brain of a twenty-week-old fetus can no more think than can a bin of unwired semiconductor chips perform like a computer. In the computer, as in the brain, the wiring has to happen first. Each cell in a fully developed human brain is in communication with tens of thousands of other brain cells. It is this richness of interconnection that makes feelings and thoughts possible.

The wiring process depends, like so many other events in the body, on apoptosis, or programmed cell death. What happens is that a great many nerve cells are formed and send out tentacles that are supposed to make contact with other nerve cells, muscles, or sensory organs. Like bloodhounds, the tentacles follow a scent trail, picking up chemical signals emitted by the cells they are supposed to reach. These are the particular "social control" signals that neurons are specialized to detect and without which they cannot long survive. So, if there is no scent trail or if the nerve goes astray, it loses the signal and dies. Neurons that link up to their proper target recieve a steady flow of life-sustaining signals. In studies of the developing optic nerve, for example, researchers have found that 10,000 nerve cells die every day. By some estimates, more than half the brain cells that arise during prenatal life and early childhood die.

This is not to be regretted. This is how the developing body ensures a proper match between the number of neurons and the number of target cells. Also, it helps ensure that only the right neurons hook up where they are supposed to. The brain cells that are to send filaments to the eyes, for example, are drawn only by chemical signals emitted by cells in the embryo eye. If they mistakenly blunder into, say, the ear, they lose their signal and die.

Even though the first synapses begin to form within the human cortex around the twentieth week, the cortex itself remains cut off from the rest of the brain—from which it must get the signals that relay sensory information—until about the twenty-second week. This is when nerve fibers from a part of the brain called the thalamus, situated deep in the center of the brain, begin to snake outward to penetrate the cortex. The thalamus is the brain's central switchboard, integrating and relaying signals from the sensory nerves to the cortex and processing commands from the cortex before they are sent to the motor nerves. A fully wired-up cortex without a thalamus is like a fully wired-up computer with no keyboard or modem to give it something to process.

Then fibers from the thalamus must make synapses with the nerve cells in the cortex, which does not happen until about the twenty-fifth week after fertilization—an age at which extremely premature babies do survive with intensive care. Even then it takes a minimum of another five weeks before the actual use of the linkups begins to resemble that of the born, at least as measured by an electroencephalograph, or brain wave, machine. Only in the thirtieth week do the newly flickering signals in the fetal brain begin to show the patterns of behavior typical of a fully developed brain. In other words, for the first six months of gestation the developing fetus is physically incapable of having any kind of conscious sensation or thought or any kind of inner experience of a sort that born humans would recognize. And only in the seventh month does it become almost inescapable that the fetus is able to feel and, in some sense, think.

Could a younger fetus still have some kind of inner experience based not on the cortex but on the more primitive [parts of the brain—the brain stem or spinal cord, for example—which do develop earlier? Nobody knows and it is not obvious how one would go about finding out.

Whatever the status of the fetus as it develops, the transition to life outside the womb is a fundamental one. Beginning weeks before birth and continuing in the first few weeks and months after birth many changes take place at the molecular level. In many cases, genes that served the fetus gradually shut down and previously dormant genes, whose products are better suited to life outside the uterus, are activated.

Among the best known are the hemoglobin genes. As was described in Chapter 5, the fetal form of hemoglobin must have special properties to take up oxygen in the placenta, where the oxygen is held by the mother's adult form of hemoglobin. In a sense fetal hemoglobin has to be stronger than adult hemoglobin to take away its oxygen. Once the baby is getting oxygen from its lungs, the weaker adult hemoglobin will do. If adults kept using fetal hemoglobin, their fetuses would be unable to get oxygen. Just as the placenta is the equivalent of a fetal lung, so it is also fetus's functional substitute for kidneys (removing waste products from the bloodstream), for its intestines (taking nutrients from the mother's bloodstream), and for its liver (producing, among other things, digestive enzymes that can process the nutrients extracted from the mother's blood into a form useful to the fetus). Upon birth, all of these various organs within the fetus must begin to function in a way that they could not before birth.

As the fetus reaches full development, it signals the uterus that it is ready to come out. Not all the details of the signaling system are known, but two key steps are at least partly understood. Both the uterus and a structure on the ovary—the corpus luteum, formed from the follicle that released the ovum that was fertilized—produce a substance called relaxin. The name was given for its chief function, to make the cervix, the tough, fibrous mouth of the uterus, soften. Throughout the pregnancy the cervix remains tightly closed, helping to keep the fetus from slipping out. At birth, however, relaxin breaks down the thick web of collagen fibers that made the cervix tough. Obstetricians speak of the cervix "ripening." Once softened, the cervix may easily be pushed open.

The other key event is the release of another protein called oxytocin which stimulates the muscles of the uterus to contract powerfully. Oxytocin is produced by the mother's hypothalamus, a multipurpose part of the brain, and stored in the pituitary gland, which is in the brain. Nerve endings in the cervix and the rest of the uterus somehow sense that the pregnancy is complete and send a signal to the brain, causing the release of oxytocin into the bloodstream. The powerful uterine muscles— sometimes said to be the most powerful muscles in the body—contract and the fetus is pushed through the softened cervix.

Within the newborn baby even more dramatic changes take place at this epochal transition. Oxygen and nutrients, which once came through the umbilical cord, must now be obtained from the lungs and the intestines, neither of which has ever functioned in this capacity before. The circulatory system that once pumped blood out to the placenta and back, bypassing the lungs and much of the liver, must now stop doing that and instead change the plumbing—shutting off arteries and veins here and opening them there. The process begins immediately as the two umbilical arteries constrict, preventing more of the baby's blood from being pumped toward the now useless placenta. A minute or so later, the one umbilical vein also squeezes shut, having waited for most of the baby's blood that was outside its body to be pumped back in. A jellylike substance surrounding the unbilical cord quickly shrinks when exposed to air, pinching off the circulation permanently. Within days the cord simply drops off.

As the newborn's chest muscles work, fluid is expelled from the lungs and air is pulled in to inflate the tiny sacs inside it, within walls so thin that gases can easily cross in and out of the bloodstream. The first effect of the inrushing oxygen is to trigger lung cells to release a protein called bradykinin, which orders the constriction of sphincter muscles that lie in a ring around certain blood vessels. The job of these muscles is to contract once and stay contracted. One squeezes shut a vessel that had allowed most blood to bypass the liver, sending it instead to the placenta, which performed liverlike duties. As that muscle constricts, blood flow is diverted to the liver, which, among many other duties, removes some unwanted substances from the blood. Gradually fibroblasts from the walls of the squeezed vessel proliferate, knitting a fibrous, permanent seal.

Other sphincters constrict blood vessels more slowly, taking a few weeks to close off circulation altogether. There is also a hole between the two upper chambers of the heart that allows the fetal lungs to receive oxygenated blood. After birth the lungs will be the source of oxygenated blood, but until then, they need oxygen from the placenta. When the

heart

sphincters reroute the blood flow inside the newborn, the difference in blood pressure between the two sides reverses. The effect is to push a flap of tissue over the hole, closing it off. In time the flap grows permanently in place, separating the two chambers. Sometimes babies are born with a flap too small to cover the hole. As a result poorly oxygenated blood flows out to the body, turning the baby a slightly blue color. These are the "blue babies" said to have "a hole in the heart." Surgical repair, though delicate, is essentially a plumbing job, finishing up one of the last of thousands of steps in the construction of a human being.

Actually, that is an overstatement. Even nine months are not enough to produce a fully developed human being. Many organ systems do not become well developed until months or years after birth. For one thing, the nervous system remains particularly undeveloped; nerve fibers running throughout the body still lack the myelin sheaths that insulate them and allow signals to be conducted properly. Even the brain continues growing in size and complexity of its internal wiring patterns for several years.

For that matter, as far as the species is concerned, human development is not complete until individuals have fully developed reproductive systems, which commonly takes at least twelve more years. Only then can they serve the essential biological function of passing on to a new generation the sparks of life carried in the sperm and egg.

Read this first paragraph and then put down this book and try the experiment it suggests. Look at the inside of your wrist. Now clench and unclench your fingers repeatedly. You should be able to see the skin rise in places and fall in others. If you unroll your sleeve and look farther up the arm near the elbow while doing the same exercise, you can probably see motions there also.

It happens all the time, but until we stop to examine how our bodies move, we take for granted one of the most remarkable properties of life—living motion. This time the motion you feel is not of microscopic vesicles gliding inside cells or of cells crawling about a dish. This is the most dramatic form of animation that life can show, the surest sign that matter has organized itself into a most remarkable mechanism.

Most of the muscles that operate your fingers, as you probably noticed in the fist-clenching exercise, are not in your fingers; they are in your forearm, in the part nearer your elbow. As you clench your fingers, you should be able to feel the muscles go from a soft, almost gelatinous mass to something hard. Each muscle is linked to its finger by a tendon, a long bundle of the fibrous protein molecules called collagen (which was secreted into this form by special, roaming cells called fibroblasts). Each muscle is like a gang of people pulling on the end of a rope, and the tendon is the rope connecting the pulling force to the load. When you clench your fist, the muscles in the arm contract, pulling the tendons that run through your wrist and into your fingers. The usually slack tendons snap into tautness, pushing other objects around inside your wrist.

Skeletal muscles—the ones that pull on bones so that you can move—are, of course, not the only muscles in the body. There are the muscles that contract just once at birth and there is the heart muscle, built to keep pumping all your life. There are also muscles in your blood vessels that squeeze to push the blood along. Humans can talk because of muscles that work the voice box and tongue and, of course, the muscles that force air out of the lungs. There are also special lung muscles that move air around inside the lung. There are muscles in your kidneys that squirt fluids from one chamber to another. The stomach has muscles to mash the food and mix it with digestive juices. The intestines have muscles to keep the digested matter moving. The uterus has muscles to push out the baby at birth. There is even a tiny muscle attached to each hair that makes it stick up when you're cold—a vestigial effort to fluff up the fur for better insulation. The same tiny muscles are in follicles with no hair; they make goose bumps. Some other muscles also respond to the cold by shivering, an attempt not to create motion but to burn energy for the heat that is released in the process. The eyes are well endowed. There are muscles that blink your eyelids, muscles that aim the eyes, and muscles that pull on the lens to change its focus. The muscle that closes the iris is a belt that encircles the pupil. When the muscle contracts, the pupil shrinks. To open it up again, other muscles positioned like spokes from the pupil out to the edge pull, and the pupil widens.

In all these muscles, though classified by anatomists into different types, the fundamental mechanism that produces the motion—the phenomenon that generates the force—is the same. More profoundly, the molecules and mechanisms that operate in muscle cells are exactly the same as those that operate in every other kind of living cell. The mechanism is the same as the one that makes a dividing cell pinch in two. In fact, that mechanism—a belt of filamentous molecules that cinch tight around the dividing cell—is essentially a muscle within a cell. And the mechanism is basically the same as the one that gives motile cells the ability to crawl around.

Thus muscles provide one more example of how evolution has adapted the same basic molecules to perform many tasks. That original process had to have arisen billions of years ago because it is virtually identical in a human cell and a protozoan cell, species that descended from evolutionary branches that must have separated billions of years ago. The chief difference between the mechanism that makes a cell divide and the mechanism that flexes fingers is that muscle cells have a higher concentration of force-generating molecules and they are arranged in a specialized architecture. Also, in the muscle cell the architecture is permanent, while in the dividing cell, it assembles only when needed and then is dismantled.

The generator of force in a muscle cell is a molecule very much like the motor molecules, such as the kinesin described in Chapter 2, that transport vesicles within cells. The mechanism is also similar to the one that makes cilia and flagella wave—the action of the motor molecule dynein as it pulls, "hand over hand," on microtubules. Inside the muscle cell there is a third kind of motor molecule, called myosin. Each myosin molecule has a long tail, making it look like a golf club. The head of the club is the part of myosin that does the work, flexing and swiveling. Myosin's tail is the club's handle (Figures 9.1 and 9.2).

Inside each muscle cell the long tails of the myosin molecules are, in

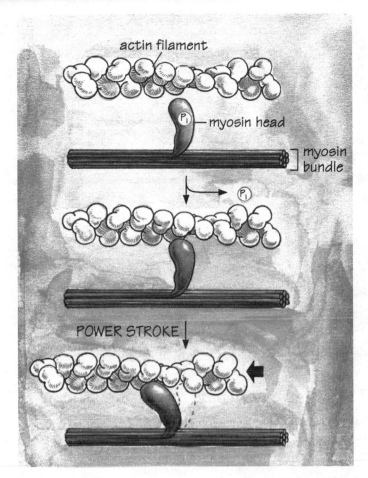

**Figure 9.1**  Within muscle cells are countless parallel filaments of long proteins called actin and myosin. Each bundle of myosins is surrounded by actin filaments, much as shown in Figure 9.2. When the proper molecular signals come along, the myosin heads projecting from the myosin bundles bind to an actin filament, flex so as to pull the myosin past the actin, release, and move back to grab again.

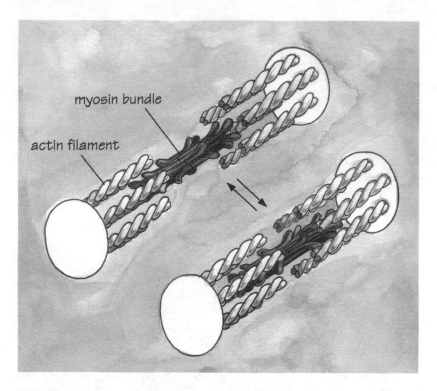

**Figure 9.2** The functional unit of a muscle is this structure—a myosin bundle surrounded by actin filaments. To the left, the muscle is in the relaxed state. When myosin heads grab actin and pull, each set of actin filaments slides toward the other, shortening the muscle. The discs represent places where other sets like these are attached end to end to make a muscle fiber that can be of great length.

effect, twined together to form a much longer bundle with myosin heads sticking out in all directions along the length of the bundle, rather like a bunch of garlic heads braided together. Lying parallel to the myosin bundle are filaments of actin, the long polymer that enables individual cells to crawl and a key component of the belt that makes cells divide. In fact, overlapping each myosin filament are six actin filaments. When a muscle contracts, each myosin head reaches out, binds to the nearest actin filament, flexes so as to pull on the actin, and causes it to slide with respect to the myosin filament. Then the myosin releases, reaches ahead to grab, and pulls again. Because many myosin heads are doing this, not in unison but with random timing, there are always some myosin heads holding the actin filament. This way the myosin bundle maintains a continuous pull on the actin filament. The result is that the two chief components of muscle cells, parallel filaments of actin and myosin, slide past one another. One way to visualize the process is to think of two or three people in a canoe floating parallel to the edge of a pier. Each person reaches to the pier and randomly grabs onto it and pulls—all in the same direction. Though some hands are releasing their grip and reaching ahead, others maintain the link and pull. Although it is always connected to the pier by some hands, the canoe moves along it.

In muscle, as might be expected, this activity is powered by the energy stored in *ATP* or *adenosine triphosphate.* Muscle cells are jammed with mitochondria, the organelles that extract food energy to make ATP.

In the human embryo the first cells to become specialized for use as muscle begin to form toward the end of the third week after fertilization, when the embryo is the size of a sesame seed. Muscle is one of the earliest adultlike tissues to appear because it is needed to operate the embryo's primitive, tubular heart. Without blood circulation, neither nutrients nor oxygen could reach all cells of the rapidly enlarging embryo.

Though the early muscle cells are destined to form only a primitive heart, their inner workings are fundamentally the same as those of all adult muscle cells, whether the muscle makes a goose bump or a high jump. The primitive heart arises as certain undifferentiated cells begin to elongate and to manufacture lots of actin and myosin filaments and to arrange them in long, overlapping bundles. The filaments are positioned so that every myosin head can reach from its bundle to any of six much thinner actin filaments around it. And every actin filament is within reach of three of the thick myosin bundles. In cross section, the arrangement is so regular in spacing that it looks like a crystal lattice, or like the precise hexagonal pattern of a honeycomb. At every corner of a hexagon there is an actin filament, and in the center of every hexagon there is a myosin filament.

Although the primitive heart muscle cell grows long and spindly, the two kinds of filaments inside do not reach from end to end. Instead they are arranged in segments. The architecture is something like that of two hair brushes arranged so that the bristle sides are facing each other but not touching. Both sets of bristles are made of actin. Floating between them and reaching into each set of bristles are the myosin filaments. If the myosin filaments could grab the actin bristles, they would pull the two brushes together.

The "brushes" inside a muscle cell are a little different. Each muscle brush has actin bristles sticking out in opposite directions, to the left and to the right. When the myosin heads grab and pull on the actins, they draw each brush toward the other. Because the heads repeatedly grab and pull, all of them together can make the entire length of muscle contract considerably. When the embryonic heart muscle contracts this way, it gets shorter and thicker, squeezing the blood vessel it is wrapped around. As the muscle develops, more and more similar cells take up positions side by side and end to end.

During the fourth week of development, skeletal muscles begin to form too, but they take a different form. Instead of each cell remaining a discrete unit, hundreds of these muscle cells give up their individuality, fusing their plasma membranes to make one gigantically long cell with hundreds of nuclei. This supercell, which is called a muscle fiber, may be several times the diameter of an average cell and a few inches long. Within this cell the actin and myosin of all the precursor cells cooperate to form many long bundles of actin and myosin, many thousands of double-sided hair brushes meshed, bristle to bristle, into a great length. When many of these supercell fibers bundle in parallel, they constitute a skeletal muscle.

The muscle is tapered at each end so that the pulling force of countless millions of myosin heads is concentrated on the tendon, which links the tapered end of the muscle to a point on a bone. The muscle is like millions of people all pulling on hundreds of ropes all spliced together at one end. The "rope" in muscles is mainly collagen. Muscle cells stick tightly to collagen, and collagen itself is strong, as anyone can attest who has tried to tear a raw chicken leg away from the carcass. Despite collagen's strength, people sometimes do "pull a muscle" or tear a tendon. When this happens, the muscle cells are literally ripped away from their attachment to the collagen or the collagen itself is torn. Fortunately, fibroblast cells are always lurking nearby, and they gradually repair the damage by secreting new collagen, in effect cementing the broken ends together. Leg and arm muscles develop from cells that arise within the embryo's torso and migrate out into the budding limbs late in the fourth week. During the fifth week they take up positions alongside other cells that are beginning to lay down cartilage, the tough gristle (which contains collagen and other fibrous proteins) that is the first step of embryonic bone formation. As the cartilage is replaced by bone and as the bones grow, the muscle cells grow with them.

In addition to heart muscle and skeletal muscle, the embryo produces a third kind, called smooth muscle. It lines the intestinal tract, the uterus, blood vessels, the lungs, and makes hairs stand on end. Like heart muscle, smooth muscle remains as discrete cells, but unlike it and unlike skeletal muscle, its actin and myosin are not arranged in the long bundles of interlocking hairbrushes. Instead, they crisscross the cell in various directions so that, when triggered, the flattish cell simply pulls into a blob. Because smooth muscle cells bind strongly to one another at the edges, they can form a sheet covering other tissue. When the cells pull themselves up into blobs, they squeeze the tissue that they surround.

The three types of muscle are controlled in three different ways. Skeletal muscle is, of course, under voluntary control. You choose when to flex it. Heart muscle operates under its own command. It contracts rhythmically all by itself, governed by a pacemaker within the heart that coordinates the entire organ, although nerve impulses from the brain can speed up or slow down the rate. But even without a pacemaker, the individual cells will contract rhythmically. In fact, heart muscle cells in culture keep right on twitching and will even join to form larger structures that "pump" uselessly in a dish. The cells do not form anything that looks like a heart, but it is conceivable, some researchers say, that a way could be found to encourage the cells to grow into a one-chambered pump that could be transplanted into people with failing hearts. If the original cells were taken from the prospective patient's heart, there should be no problem with rejection by the immune system. The third muscle type, the smooth muscle that moves many internal organs, is under control of the autonomic nervous system, which is operated by the unconscious part of the brain. Some muscles, such as those of breathing, can be controlled consciously and unconsciously.

By the time of birth or within a few months after, a person has essentially all the muscle cells he or she will ever have. Once it becomes fully specialized, a muscle cell is so full of the machinery for generating a force that it can no longer divide. However, if skeletal muscle cells are destroyed because of injury, new cells can be produced by certain embryolike cells that wait in dormancy for just such occasions. These undifferentiated cells lurk in compartments adjacent to muscle fibers and are triggered into dividing by the events of the injury. (The next chapter is on wound healing.) While these cells can restore considerable muscle function, a severely damaged muscle will never again be as strong as normal.

Muscle cells do, however, get bigger as the person grows and, of course, with exercise. As many physical fitness buffs know, there are two fundamental ways to exercise, and each produces a different result. Low-intensity, long-duration exercise such as distance running and swimming builds up endurance without much gain in the power a muscle can exert in a single flex. On the other hand, high-intensity, short-duration exercise, such as weight lifting, builds power but doesn't add to a person's endurance. Both kinds of exercise have their effects at the cellular level because there are three basic types of skeletal muscle fibers (the supercells) and they get and use their energy in different combinations of ways.

In general, endurance-building exercise causes muscle cells to produce more mitochondria, to make ATP. And exercise encourages the growth of more blood vessels into the muscle, to bring in oxygen and nutrients and to carry away waste products more effectively. Power-building exercise, by contrast, causes muscle cells to manufacture more filaments of actin and myosin and to install them next to the existing ones within each cell. In weight-training this buildup happens if the muscles are worked at more than 40 percent of capacity, making muscles thicker. The nature of the mechanism by which exercise tells the muscle cells to make more filaments, however, is still a mystery. Endurance training, on the other hand, can make muscles slightly smaller, even as it makes them better able to work for long periods.

The effects of exercise vary from one muscle (bundle of muscle fibers, or muscle cells) to the next because each muscle is composed of three different kinds of cells in different proportions. Some muscles are rich in cells that contract slowly (they contain a form of myosin that responds more slowly to ATP) but are resistant to fatigue—the muscles of the back, for example, which are involved in maintaining erect posture. These endurance cells, which get their ATP from mitochondria, are also rich in a protein called myoglobin, which is like the blood's hemoglobin in that it can take up oxygen, hold it, and release it on demand. Like hemoglobin, myoglobin is red, and so muscle with a preponderance of these cells is dark. The dark meat of a chicken, for instance, is rich in long-endurance muscle cells.

Other muscles are rich in cells with a form of myosin that contracts rapidly, producing a strong, brief burst of power. The upper arms are rich in such cells. They get their ATP primarily from the breakdown of glycogen, a carbohydrate molecule made of many sugar molecules and stored in the cells. This process requires no immediate source of oxygen, and the cells that do it lack myoglobin. As a result the muscle is light, like the white meat of chicken.

There is a third form of muscle cell that mixes attributes of the other two kinds. It can flex quickly but is intermediate in ability to function without fatigue. Depending on the type of exercise, the appropriate muscle cells will receive the benefit.

Without regular, moderate exercise, all muscle cells will dismantle many of their actin and myosin filaments and the numbers of mitochondria drop. The muscle wastes away, or atrophies. Also, the muscle's richer blood supply disintegrates. In most cases, however, this muscle atrophy can be reversed if exercise resumes.

How does a muscle know when to contract? An early clue emerged in 1791 when the Italian physician Luigi Galvani was dissecting the muscular hind leg of a frog and suddenly the leg jerked as if alive. But the rest of the frog was absent. Galvani, whose name is the root of our word "galvanized," thought he had discovered a new form of electricity produced in muscle. A few years later his friend, the physicist Allessandro Volta, found a better explanation, one that led him to invent the electric battery. Volta showed that Galvani's scalpel was touching a wet piece of brass when it also touched the frog's leg. The two metals had acted like a battery and produced a tiny jolt of electric current. The muscle didn't make electricity, Volta said. The two metals made the electricity, and the current made the muscle contract.

Volta was not quite correct. All living cells do, in a sense, make electricity, and the body uses electricity to make muscles work. To understand this, one must first know a little about the electricity that is so much a part of the life of many cells. Most cells are surrounded by an extracellular fluid that contains mostly positively charged atoms, or positive ions. Because opposite charges attract, this means that atoms and molecules with negative charges inside the cell will be drawn as close as they can get to the opposite charge. Usually this means they will cluster at the inner face of the cell membrane. If the charged molecules and atoms could get across the membrane, they would combine. The avidity with which they "try" to move together has the potential to do work and is measured in volts, the same volts (named for Allessandro Volta, whom we just met) that come out of a battery or a wall outlet. The two poles of a battery, positive and negative, are comparable to the two sides of a cell membrane. Of course, in cells the voltage is tiny, measured in thousandths of volts.

But it is powerful enough to use in the body as the signal that commands skeletal muscles to contract. The current starts in the central nervous system, which consists of the brain and the spinal cord. If you did the experiment suggested at the beginning of this chapter, you made a conscious decision to flex your fingers. An electrical impulse traveled from your brain into your spinal cord and then out through the so-called motor nerves at the base of the neck into the muscles in your forearm. The "cell body" of every voluntary motor nerve lies within the spinal cord or in the brain stem just inside the skull. From each of these cell bodies a filament, the axon, reaches out a long distance, sometimes several feet. The tip of each motor nerve reaches into a muscle and then sprouts branches, anywhere from about a dozen to a hundred. The tip of each branch reaches out and almost touches one nerve fiber, one supercell, usually in the middle of its length at a place called the neuromuscular junction.

This extraordinary structure, of which the body has many thousands, is the transition point, the link between the two great human realms with which philosophers have grappled for millennia—thought and action—or, if you prefer a metaphysical view, it is the link between the spirit and the flesh. The neuromuscular junction is as far as the brain and nervous system can reach, and if the junction does its job correctly, it is the origin of behavior. When a signal comes along a motor nerve axon, it splits repeatedly to move through all the axon's branches, triggering all the fibers to which the branches are linked. Within a given muscle there may be many axons, each one triggering its own group of fibers. This is a kind of insurance against losing the entire muscle if a single nerve were to die.

The nerve signal is a tiny pulse of electrical charge, but when the pulse reaches the tip of the axon's branches—where it almost touches the muscle fibers—the situation changes. The pulse causes calcium channels in the nerve cell's membrane to open. In transmuting thought into action, calcium ions (calcium atoms with an electrical charge) play many roles.

Indeed, calcium ions also play key roles in many other cellular processes, shuttling signals about because of their inherent electrical charge. In this case when the calcium channel pumps calcium ions into the axon, the ions cause vesicles filled with the neurotransmitter acetylcholine, to dump their contents outside the cell. This is the end of the nervous system's role. At this moment the powers of thought have no further ability to function. The released neurotransmitters drift across the gap between nerve and muscle and bind to receptors on the muscle cell. This binding triggers a new electrical impulse that moves over the membrane covering the muscle fiber, spreading in both directions. The electrical impulse causes a reaction inside the muscle fiber: Still more calcium ions, stored in compartments in the muscle cell, are released into the muscle cell's interior.

This calcium, in turn, binds to proteins that stick to the actin, covering its sites to which myosin can bind. When the calcium binds to these covering proteins, their shape changes so that they are pulled out of position, exposing the myosin-binding sites. The myosin heads spontaneously grab the actin filaments and flex, pulling on the actin. It seems like a clumsy, Rube Goldberg contraption but the time between the arrival of a nerve impulse at the neuromuscular junction and the contraction of the muscle is a mere ten milliseconds, or ten one-thousandths of a second. If it took much longer, tennis would be a very slow game.

Calcium's crucial role in this process, incidentally, has led to a useful new treatment for people with heart disease. With their condition, their hearts beat too fast, especially in response to stress. Many such patients now, however, take a drug that blocks their calcium channels. The channel blocker cuts down on the calcium released within the heart muscle, making the organ beat with less force. This reduces high blood pressure and lowers the heart's need for oxygen, making it easier to get along on a blood supply from arteries narrowed by disease such as atherosclerosis. Of course, as with virtually all drugs, there is a side effect. In this case the drug acts on all other muscles as well, reducing their power.

As well as it works, the neuromuscular junction is the site of many troubles.

Myasthenia gravis, a progressive disease that leads to severe, permanent muscle fatigue and weakness in about 100,000 Americans, is the result of a person's immune system mounting a bizarre attack on the junction's acetylcholine receptors. For reasons that are not yet understood, the immune system loses its ability to distinguish foreign proteins from certain native ones and makes antibodies that bind to the protein that functions as the receptor, destroying it. As the number of receptors declines over a period of years, nerve signals reaching the muscles become fainter and the muscle cannot contract as strongly. Paralysis gradually gets worse.

Deep in the rain forests of the Amazon, several tribes have discovered a faster way to block the action of acetylcholine at the neuromuscular junction and thus cause rapid paralysis. They make poison arrows, dipped in the juice of certain plants. An animal struck by the arrow quickly becomes paralyzed. Medical researchers call the poison curare, from its native name, and have learned that the substance finds its way through the bloodstream to the neuromuscular junction and binds strongly to the acetylcholine receptor but does not trigger the receptor to activate the muscle. This means that when a nerve signal reaches the muscle, there are no available receptors for acetylcholine. The muscle is paralyzed. Curare and similar drugs are used in some surgery to keep muscles from twitching during the operation. Then they are removed with other drugs that break the curare down.

In normal neuromuscular junctions, spent acetylcholine is degraded by a special enzyme. This frees the muscle's receptors for another signal. Nerve gas, one of the agents of chemical warfare, destroys the enzyme that does this, leaving the muscle unable to reset itself for another signal, and again, the muscle becomes paralyzed. Yet another cause of paralysis at the neuromuscular junction is botulism, the most deadly form of food poisoning. Certain bacteria, called *Clostridium botulinum,* secrete a protein that blocks the release of acetylcholine from the nerve ending. The brain can send a signal, but the signal never emerges from the nerve. Polio, which reached epidemic proportions in the United States in the 1950s, also paralyzed by depriving muscles of the nerve signals they need to contract but through a permanent action. The virus killed the nerves outright. Without a signal, the muscles could not contract and withered from disuse.

Smooth muscle, the kind that operates in many internal organs and that is under command of the unconscious part of the brain, is controlled in several different ways. Since smooth muscle retains its discrete cells and does not fuse into supercells, nerve endings must either be in contact with many muscle cells or the cells must be linked by gap junctions, the tiny cell-to-cell pipes through which electrically charged calcium ions may pass to relay the nerve's signal. Both methods exist, depending on the muscle. Smooth muscle cells can also be triggered by certain hormones. For example, the uterus, as we saw earlier, is stimulated to contract by hormones from the fetus. Also, there are two kinds of nerves that command smooth muscle, one that stimulates it to contract and another that blocks its reaction to other triggers such as hormones.

The cells of the heart, the third type of muscle in the body, are also separate (no supercells), but they are linked by gap junctions that allow calcium ions to pass, and as we have seen, calcium ions trigger the contraction of actin and myosin. As was mentioned before, no nerves tell the heart muscle to contract, although they do vary the rate of beating. Control of the heartbeat is by a dime-sized patch of cells near the top of the heart, the pacemaker. Like most cells, they are under the influence of channels that control the flow of electrically charged atoms into and out of the cell.

The electrical charge carried by many cells persists a long time while the cell goes about its business. Particles that are positively charged, such as sodium ions, may come into the cell through various channels, but the cell has structures in the membrane that pump them back out and pump in compensating particles of opposite charge. Cells, by and large, keep unlike charges fairly well separated on either side of their outer membranes. This maintains the cell in what is called a polarized state. But certain events can upset this, triggering dramatic events within the cell. Not only can sudden depolarization lead muscle cells to contract, as we have seen, but the same mechanism triggers unfertilized eggs to begin dividing as if a sperm had penetrated them.

The cells of the heart's pacemaker, however, are different. Their normal condition is not to maintain the usual steady polarization but to let themselves slowly depolarize. This happens because positively charged sodium ions trickle into the cell at a slow rate and the pacemaker cells, unlike other cells, do not immediately balance this by pumping sodium ions out. Instead, the cell declines in polarity until it reaches a lower threshold. Then, suddenly, the ion pumps kick in to restore the original polarity. This sudden burst is the heart's equivalent of an electrical impulse in a nerve. Special heart cells, shaped into filaments that act like nerves, conduct the impulse to other parts of the heart, inducing them to beat. The electrocardiogram (ECG or EKG) is a way of studying how well this signal travels to all parts of the heart. Electrodes touching the skin at various points pick up the electromagnetic waves that emanate from the traveling impulses, much as a radio antenna picks up radio waves given off by a current flowing in a transmitter antenna.

The role of electricity and ions in muscle function is of more than academic interest. It is also why there is Gatorade and other so-called sports drinks. Athletes who work hard lose ions through sweating, mainly sodium and potassium. Also, if they become dehydrated, the balance of ions in the bloodstream, often called fluid electrolytes, can become upset. As a result, muscles can fail to work at their best. Even nonathletes who get dehydrated risk losing muscle function as a result of electrolyte imbalance. Muscles can go into spasm or start twitching crazily. Nerve signals may fail to stimulate muscles. Muscles may be stimulated even without a nerve signal. Much of the trouble is caused by the levels of calcium falling too low in the extracellular fluid. One result is cramps. The electrical impulse that spreads along the muscle may fire as often as 300 times a second, forcing the muscle into a painful, essentially continuous effort to contract. Gatorade is, of course, a soft drink favored by athletes because it contains significant doses of sodium, calcium, and potassium ions that are supposed to make up for the losses during sporting events. The drink's value, however, is based more on reputation than data. Studies of athletic performance show that Gatorade's most valuable ingredient is the water. It seems that the biggest contributor to electrolyte imbalance in healthy people is not the loss of ions but the loss of water through sweating. Prevent dehydration and there are plenty of reserve electrolytes in the body to keep their concentrations at good levels. A normal balanced diet more than makes up for any actual loss and provides enough reserves to get most people through considerable periods of electrolyte loss.

One of the most serious conditions causing loss of muscle function is muscular dystrophy. It is one of the most common serious inherited diseases in the United States. About one in every thirty-five hundred boys born in this country has the disease, which begins to appear as muscular weakness between the ages of three and five. By the age of twelve, most victims are slumped in wheelchairs, the spreading paralysis dooming them to die in their teens. Few survive beyond their twenties. There is no specific treatment.

After generations of suffering and decades of biomedical puzzlement, however, the conquest of muscular dystrophy may be at hand. The explanation of muscular dystrophy at the cellular and molecular levels represents one of the major achievements of modern biomedical science. More than that, it is an achievement that is pointing toward a possible treatment or even, someday, a cure.

The fundamental cause of muscular dystrophy, a defective gene, was found in the late 1980s. Because muscular dystrophy was almost exclusively a disease of males, researchers concluded long ago that it had to be a result of a defect in a gene carried on the X chromosome. Females have two X chromosomes, so if the gene on one is bad, the odds are overwhelming that the other is good. Since there are two copies of most genes in the body (one on each member of a chromosome pair), there is usually no problem if one is defective because the other can handle the job alone. Males, on the other hand, have only one X chromosome. If the gene on it is faulty, there is no backup.

Louis M. Kunkel and his colleagues at Harvard Medical School have found the muscular dystrophy gene, a step that has already made it possible to diagnose the disease precisely in early life, a task that had long been difficult because the early symptoms are similar to those of many other muscular disorders. The researchers are hot on the trail of establishing the gene product's function within muscle cells and are even beginning to imagine treatments for the as-yet incurable disease.

The gene has turned out to be the biggest one ever found. It is 2.2 million bases long and consumes an astonishing 0.1 percent of the entire coding potential of the human genome. If every other gene was as big, the human genome could contain only 1,000 genes. In reality, it has between 50,000 and 100,000. Although most of the gene consists of intervening spacer sequences, the resulting protein is still one of the largest known, consisting of nearly 4,700 amino acids. Kunkel has named the protein dystrophin.

After examining the genes of hundreds of muscular dystrophy patients, the researchers have found that all sufferers have mutations that have removed one or more of the exons, or coding segments, of the dystrophin gene. As it happens there are two ways exons can be deleted from a gene, and they correspond perfectly to the two types of muscular dystrophy—Duchenne muscular dystrophy, which is the more severe and more common kind, and Becker muscular dystrophy, which is milder.

People with Duchenne muscular dystrophy, the researchers have found, have a dystrophin gene in which the deletion causes a shift in the reading frame, a disastrous kind of mutation because it yields a completely useless protein. As we saw earlier, genes use a sequence of three bases to specify each amino acid. These three-base codons are immediately adjacent to one another. Thus there is no "punctuation" between codons. This means that the ribosome, as it is reading the messenger RNA to make the protein, simply interprets the next group of three as specifying a particular amino acid. Each triplet codon makes sense only if the ribosome uses the right reading frame. In other words, the sentence, "Thefatcatsat" makes sense if you know the right reading frame, which in this case is a three-letter codon starting with the first letter. But if the first codon is mutated to one letter, it would shift the reading frame and be read as: "Afa tca tsa t," which makes no sense. Duchenne muscular dystrophy victims have inherited just such a frame-shifting mutation. The ribosome assembles a completely wrong sequence of amino acids that could never do dystrophin's job. People with this form of the disease have no dystrophin in their cells.

Those suffering from the milder form have a different mutation, one that preserves the correct reading frame because it happens to delete exactly one triplet codon. The hypothetical sentence above, with a frame-saving deletion, might read, "The cat sat," which still makes sense although it loses some of the meaning. People with Becker muscular dystrophy have such a mutation in their dystrophin gene, and as a result, their ribosomes make a dystrophin protein that is deficient in one way or another and that cannot do its job perfectly. These people have varying degrees of mild-to-moderate muscle dysfunction, the degree depending on where the dystrophin molecule is missing a key part.

What is dystrophin's job? The full story has not been worked out, but it is clear that dystrophin is a close relative of other, better-known proteins that link the cytoskeleton to the cell's outer membrane. Dystrophin resides just inside the muscle cell's outer membrane—where the signal to contract is received and propagated throughout the muscle fiber—linking it to the contractile machinery deeper inside. Its role is obviously crucial because without the proper form, muscles are dramatically weakened or even useless.

(Incidentally, dystrophin is also expressed in a very different part of the body—the brain—as well as at specific sites within the neurons, the cells that do the thinking. There is evidence that dystrophin is situated just inside the synapses of the neuron on the receiving end of a nerve signal. One speculation is that the molecule is needed to anchor the so-called postsynaptic apparatus. In that capacity, its derangement could account for the neurological effects of muscular dystrophy.)

Biomedical scientists have even "cured" individual dystrophic muscles in a tantalizing experiment. They removed some of the undifferentiated muscle-repair cells from the muscles of two patients and implanted a good copy of the dystrophin gene into them. Then the cells were injected back into a muscle of each patient. Within weeks the cells had responded as if they were repairing a damaged muscle. They spawned new muscle fibers equipped with dystrophin and the once-paralyzed muscles regained some strength. Though intriguing, the approach is not seen as feasible for treating the whole body since hundreds of muscles would have to be injected many times each and even then the regained strength would not be great.

The new findings raise another question. If the total absence of dystrophin, as is the case with the Duchenne form, paralyzes the muscles, why aren't children born paralyzed? How come their muscles work at first? The answer, according to Kunkel, seems to be that during fetal development and infancy, muscles have a different protein that does the same job. But as the child ages, that gene somehow slows down its production and the dystrophin gene is supposed to take over. This shift from proteins that work mainly in fetal life to different proteins intended for adult life is true for a number of other proteins, such as hemoglobin.

If this turns out to be the case, it would raise an intriguing possibility. The fetal gene should still be present in all muscle cells. Would it be possible to reactivate it? Perhaps some drug could be found that would get into each muscle cell and turn on the fetal gene. Nobody knows now whether such a thing is possible, but it is an idea that, if successful, could lead to the conquest of a major cause of human suffering. It is also an idea, like so many that are beginning to emerge all across biomedical science, that could only come from an intimate understanding of how life works at the cellular and molecular levels.

# HEAL THYSELF

CHAPTER **10**

When the republic of cells is threatened by physical damage, a panoply of responses explodes into action. The body has a 911 Rescue Squad, a SWAT team, a Rapid Deployment Force, a Red Cross, and a few other emergency response mechanisms that spring into coordinated action. A variety of specialized cells—some that were patrolling in the bloodstream and some that were waiting patiently in a kind of dormancy while nestled among the working cells of any given tissue—take emergency measures. The rescuer cells stop the bleeding, then clean up the debris of dead, dying, and damaged cells, and finally set about repairing and replacing whatever structures were lost.

In the end, the wounded tissue is often as good as new. So reliable is the wound-healing process, in fact, that most people take it for granted that even deep cuts, severe abrasions, many burns, and broken bones will heal well. The ability of cells to repair damage is a wonder. But the body's wound-healing mechanisms are not so powerful that they can repair such major injuries as, say, the loss of an arm or an eye. The lack of an ability to regenerate these lost structures is itself also a wonder, for there are many species of lower animals in which such seemingly catastrophic wounds are only a minor setback.

If, for example, a salamander loses an entire leg to a snake (or to a curious biologist), the animal simply grows a new leg. In an unfolding sequence of cellular events that looks very much like a replay of the development of a limb during embryonic life, a bud sprouts from the amputated arm's stump and, over a period of days or weeks, gradually develops into a whole new limb complete with bones, muscles, tendons, nerves, and skin. The new limb has a proper knee or elbow joint, it has a foot with the right number of toes. Why, generations of biologists have puzzled, can't the human body do the same thing? What a boon it would be. Even if it took a year or more of slow growth, think of the many handicapped people who could benefit. The ability to regenerate major structures would be of such obvious survival value that it has long been a mystery why birds and mammals lost the ability. Surely natural selection would have forsaken such an ability only in exchange for some other ability of even greater survival value. No one knows, however, what might have been gained in the trade. As it is, the only significant forms of regeneration of which humans are capable is that which rebuilds the blood, which is often regarded as an organ in its own right, and the skin. Liver tissue is also capable of fairly significant regeneration.

The salamander is not even special in the animal kingdom. Frogs can do much the same. The powers of regeneration are even more astounding among the more primitive (in evolutionary terms) species. Some starfish, for example, can heal even more severe wounds than can amphibians. Tear one apart arm by arm (as did divers a few years ago when they were trying to protect South Pacific coral reefs being devastated by the crown-of-thorns starfish), throw the pieces back into the sea, and each piece that retains part of the central disk will regenerate the missing arms and become a whole starfish. Chop in two a flatworm called *Planaria,* an even more primitive animal that dwells in ponds and high school biology labs, and the head end will grow a new tail while the tail end grows a new head. Go still further back on the evolutionary scale to sponges—which are among the simplest multicelled animals—and an even more prodigious feat of regeneration is possible. As researchers learned early in this century, you can push a sponge through a fine sieve to separate it into individual cells and the cells will settle onto the walls of the aquarium, crawl about until they find one another, and form hundreds of clumps that are essentially minisponges. Then each clump will grow into a complete new sponge.

Go still further back, to the one-celled protozoans, and the concept of wound healing merges with that of cell division. In a sense, single cells chop themselves in half and then grow back to full size, re-creating the original complement of organelles proper to a full-sized cell. And, of course, if cells are to heal a wound in a multicellular organism, they must engage in prodigious cell division, splitting and regenerating themselves to regenerate the damaged or lost tissue.

The signals that command cells to begin the processes of wound healing are not well understood, but there is evidence that the disrupted environment somehow activates the same suites of genes that normally operate only during embryonic development. The creation of tissues and larger structures that takes place is, in a sense, a replay of what happened much earlier when the body was first forming itself. When the cellular microenvironments reach the mature stage, the rapidly dividing cells slow down and fall into lockstep with their undamaged neighbors.

In the case of the salamander's amputated leg, for example, the altered microenvironment at the stump appears to activate certain dormant cells, called mesenchyme. Mesenchyme cells are undifferentiated, much like those that during embryonic development would migrate about the body to reach various sites and then change into specialized shapes. In this process, however, not all the cells become specialized. Some remain in their amorphous state, scattered through many of the body's tissues. Mesenchymal cells are the uncommitted voters of the cellular republic, retaining a wide range of developmental options.

As long as the salamander's original leg is in place, these cells remain dormant. But once the leg is gone, the cells respond as if they are in an embryo's limb bud. They begin dividing rapidly, spinning off new cells that crawl about, encountering their own microenvironments and differentiating into all the normal cell types appropriate to a salamander's leg. Gradually, the proliferating cells construct a new leg and the microenvironments again approach that of the mature tissues. The cells shut down their developmental programs. The "grow-a-leg" set of genes switches off. Along the way, however, the old mesenchyme cells produced a few new mesenchyme cells to leave behind, scattered through the leg, to wait and monitor their microenvironments.

The body's first priority when a blood vessel is cut is to stop the bleeding. Several processes are involved. One involves a curious entity in the blood called the platelet. Platelets look like miniature cells, but they have no nuclei or other organelles. Actually, they are tiny pieces which have broken off from very large cells called megakaryocytes, which live in the bone marrow. Little bits of the fringe of a megakaryocyte pinch off and drift into the bloodstream. Countless billions of platelets circulate in the blood. Inside each platelet are lots of molecules of actin in the monomer form. This is the same molecule that individual cells use to crawl and that muscle cells use with myosin to contract. In platelets, however, the actin stays in the monomer form because each individual actin molecule is capped by a molecule of another protein called profilin.

When a blood vessel is cut, the opening spills the platelets into the extracellular space, which contains a tangled web of filaments of collagen, the tough, fibrous protein found in the extracellular matrix but not inside the blood vessel where the blood flows. Platelets carry binding sites for collagen, so as soon as the platelets touch the collagen, they stick to it. The act of binding activates the platelet, sending a signal inside that causes the profilin cap to come off each actin monomer. The actin quickly polymerizes into filaments that push outward on the platelet's membrane, stretching it into long arms that reach out like tentacles. At the same time, the platelets release a substance that activates nearby platelets, recruiting them to the job. The platelets develop a stickiness for each other. The combination of the tentacles and the stickiness become part of a chain reaction, making the platelets aggregate in growing clumps that quickly plug the leak in the blood vessel. Thus, usually in minutes, the tiny platelets grow into a large enough aggregation to put a plug in the leaking dike and hold it there long enough for slower healing mechanisms to perform their job.

Why doesn't the platelet chain reaction continue, eventually capturing all the platelets in the body and forming one vast aggregation that blocks off all circulation? As with so many other processes in the body, there is a braking mechanism to keep things under control. The cells that line the inside of the blood vessel secrete an enzyme that acts on the substance released by platelets to recruit other platelets. This enzyme converts it into a different substance, called prostaglandin, that can no longer flag down more platelets. There are several kinds of prostaglandin (named because they were first found in the prostate gland and wrongly thought to be special to that organ), and this one happens to be a powerful inhibitor of platelet aggregation. The two molecules—the platelet activator and the platelet inhibitor—are released in rather evenly competing amounts so that the platelet clump forms quite readily at the site of the wound but not away from it.

The so-called platelet plug is only a quick fix. The next step is for the blood to construct a longer-lasting barrier to bleeding. This is the process called blood coagulation, or clotting—a two-faced phenomenon that can save lives by stopping blood loss and take lives by causing heart attacks and strokes. Circulating in the blood at all times is a rod-shaped protein called fibrinogen, which is made in the liver. Most of the time fibrinogen simply drifts along, unable to stick to anything. But at the site of a wound, enzymes circulating in the blood are activated—again by contact with the protein collagen that normally lies just outside the blood vessel—to cause a cascade of chemical reactions. In other words, A acts as an enzyme causing B to change into C, which is an enzyme that turns D into E, which acts as an enzyme to make F become G. And so on for at least seven known steps. Several forms of hemophilia, incidentally, are the result of a defect in one or another of the genes for these clotting factors. At the end of the normal cascade of reactions, small parts of the fibrinogen molecule are cut away, leaving largish remnants, called fibrin, that now polymerize into long strings and ropes.

In moments the essentially liquid nature of blood thickens into a dense gel at the site of the wound. The result quickly becomes an almost solid mass of platelets, red blood cells, and other cellular inhabitants of the bloodstream, all trapped in tangled webs of fibrin. Under the microscope a red blood cell in a blood clot looks like an inner tube lashed to the family car by somebody who had many yards of rope but no plan of organization. Where the fibrin filaments cross, they form tight bonds to one another, further stabilizing the clot. Most of the time, if the wound is not too large, blood clots form a durable patch, sealing the blood vessel against further leakage of blood. Clot formation begins within minutes and may continue for a few days, gradually enlarging the barrier to bleeding.

Sometimes blood clots form not at a wound site but inside an intact blood vessel, especially one whose inner surfaces have been roughened by accumulation of cholesterol deposits. A buildup of the fatty molecules can narrow the opening through an artery so much that relatively little blood gets through and the oxygen-deprived tissues become painful. When this happens to the arteries that feed the heart muscle—called coronary arteries because they are like a crown—the pain is called angina pectoris. The greater danger, however, is the formation of a blood clot that lodges in a narrowed coronary artery opening, blocking all blood flow. This is a heart attack, or coronary thrombosis, one of the most common causes of death in the world. A portion of the heart muscle is literally killed for lack of oxygen. Sometimes the cholesterol-damaged walls of the artery trigger the enzymatic cascade leading to a clot. It can happen in the coronary artery or it can happen elsewhere and the clot can break loose and drift with the blood until it lodges in an artery too small to let it pass. This too can cause a heart attack, or if the clot gets stuck in an artery feeding the brain, the result is a stroke. Some of the newest drugs for treating heart attacks, incidentally, are the so-called clot busters, or thrombolytics. These are synthetic versions of the natural enzymes that the body makes to keep clots from growing too large, from solidifying the entire bloodstream. Several different enzymes have been developed, but all essentially speed the breakdown of the thrombin filaments holding the clot together. If the clot is dissolved before the heart suffers too much damage, the effect of the attack can be minimized.

In the case of an ordinary flesh wound, however, the clot usually stays put, growing only just large enough to cover the wound. Moreover, special cells lurking just outside the blood vessel, called fibroblasts, begin crawling over the surface of the clot, secreting fibronectin, a protein that essentially glues the fibrin filaments to surrounding tissue.

Among the first cells to begin the process of wound healing are the macrophages, those voracious denizens of bloodstream and body. They swarm over the site, gobbling up the debris of damaged and dying cells. They also engulf any bacteria that may have gotten into the wound, an early part of the immune system's efforts to protect the body against infection and the focus of the next chapter.

Now the healing process begins in earnest. Both the platelets and the macrophages secrete molecules that signal other types of cells to join the effort. These molecules belong to a large class of at least thirty known growth factors—substances that trigger one or more types of cells to begin dividing. Among the first to go into action in a wound is fibroblast growth factor, FGF. It awakens fibroblasts, some of which were already crawling over the clot, laying down fibronectin. Fibroblasts are derived from mesenchymal cells and, as such, retain much of their embryonic ability to change into various types of specialized cells. Fibroblasts can, for example, change into cells that specialize in making various kinds of connective tissue such as the tough protein collagen, cells that make cartilage, and cells that make bone. Several growth factors are known to be involved, each part of a complex, orchestrated web of messages and responses from various types of cells.

When molecules of FGF touch a fibroblast's FGF receptor, for example, a signal is relayed to the cell's nucleus, activating a special set of genes. Though the cells had been virtually dormant, they now begin to divide, making more and more fibroblasts, many of which crawl out onto the exposed surfaces of the wound, where they lay down a surface of fibronectin. The fibronectin will serve as a roadway for the next cell type to come into play.

Other growth factors reach nearby skin cells, stimulating them to join the healing effort. Skin cells, which normally maintain strong junctions with one another, such as the spot-welding desmosomes, retract their links and dismantle their desmosomes. They redeploy the actin in their cytoskeletons to form the meshwork just inside the cell membrane that helps the long-immobile cell pick itself up and move to a new location. The epidermal cells also manufacture new kinds of surface receptors, specially tailored to let them grip the coating of fibronectin that the fibroblasts have laid down. Within hours of the injury, rapidly multiplying epidermal cells are swarming out and over the wound surface. They distribute themselves in a thin sheet, one cell thick, and then settle down. They keep dividing, but now the daughter cells do not crawl away. Instead they form a new layer above the old one. Over a period of days, the tenuous sheet of skin cells grows into a thicker and thicker layer, eventually becoming as tough as before.

Depending on the extent of the injury below the skin, healing may be complete or it may continue for weeks or months. If muscle has been damaged, special mesenchymal cells in that tissue are activated to make new muscle cells. Bone, as we have seen, is continually being remodeled by cells that break it down and build it up. When a bone breaks, still more standby cells are prodded into action. Even blood vessels regrow. One of the growth factors released in wounds beckons to existing blood vessels, encouraging them to sprout buds that elongate into capillaries, snaking their way into the healing tissue so that no new cell is ever more than a few cell thicknesses away from the blood supply.

The first known growth factor was first discovered during World War II by a pioneering Italian scientist named Rita Levi-Montalcini, who had been shut out of her laboratory by the war but who continued to experiment in her bedroom at home. She studied embryos developing in chicken eggs and found a substance that is required to make nerves grow in the embryos, now called nerve growth factor or NGF. But it was not until recent years that much further progress was made in the study of growth factors and Levi-Montalcini's early work came to be seen for the pioneering advance it was. She was eventually awarded a Nobel prize.

Knowledge of most growth factors remained elusive because the substances are present in living tissues in extremely small amounts. Detecting the presence of a growth factor was hard enough. Harvesting a reasonable quantity of it from tissues to do experiments was even harder. The advent of recombinant DNA technology in recent years, however, has made it possible to put the genes for individual growth factors into bacteria and induce them to manufacture large amounts. Only since the late 1980s have researchers finally had enough of the isolated substances to consider using growth factors as drugs to speed wound healing. Clinical trials are underway to treat skin ulcers and slow-healing wounds with topically applied epidermal growth factor. Preliminary results show that wounds that festered unchanged suddenly begin healing.

One intriguing discovery is that saliva contains epidermal growth factor. This would appear to explain why many animals lick their wounds and it suggests that humans could do the same. In fact, informal experiments do show that licking superficial skin wounds makes them heal faster. The risk of infection is also minimized, researchers have found, because saliva also contains natural antibiotics, which is why injuries to the mouth almost never become infected.

Research on wound healing has raised an extraordinary new hope among some scientists—the prospect of being able to grow new human organs in the laboratory, for research and perhaps even for transplantation.

One foundation for this hope is the work on so-called organogenesis, such as the mouse mammary cells that were induced to form rudimentary milk ducts (Chapter 8). That work suggested that if the right physical environment is present—the right extracellular proteins, in that case— cells can be made to express specialized suites of genes and remodel themselves into specialized forms. Similar work has caused disaggregated liver cells to begin to form a rudimentary liver and even nerve cells to begin making synapses in culture, forming a simple nervous system. Other cells have been made to grow cartilage and bone in the lab.

The other foundation for the hope of growing organs in the lab is the work on growth factors. There are more than thirty known growth factors, each with specialized cellular targets, and more are being discovered almost monthly. Biological visionaries are already speculating that it may be possible to supply them in carefully orchestrated doses to cultured cells so that the cells "think" they are in an embryo and proceed to grow, say, a liver or even a heart. Perhaps it might even be possible to make the cells grow a new finger or arm. If the starter cells were taken from the prospective recipient, the laboratory-grown organ should be a perfect match and would pose no risk of rejection by the immune system.

The fulfillment of such a promise would, of course, be many years away, assuming no insurmountable barrier were to be discovered. If it happened, though, it would represent nothing more than the human species recapturing one of the fundamental abilities evolution long ago gave to many "lower" species, such as the flatworm and the salamander, but sacrificed somewhere along the way. After all, human cells are fundamentally no different from those of other species. Whatever happens in the flatworm to grow a new head or in the salamander to grow a new leg, might well be made to happen in the carefully controlled environment of a laboratory container of nutrient broth.

In a natural world teeming with millions of forms of life that have evolved to exploit every conceivable ecological niche, it should come as no surprise that the interior of the human body offers some of the richest habitats available. It is always warm and wet and, of course, it is richly supplied with nutritious organic compounds.

And so it is that the body is normally home to scores of species of tiny creatures—bacteria, viruses, fungi such as molds and yeasts, and sometimes even protozoans and worms. Some are beneficial, such as the bacteria that help digest food in the gut, but some can become harmful if allowed to multiply to large enough numbers. And, of course, a few microbes are dangerous whenever they successfully invade the body.

Many of humanity's worst scourges are caused by such invaders, from the bubonic plague bacteria that decimated Europe centuries ago and then mysteriously faded, to the diabolical AIDS virus still ravaging the species. There are the more routine microbial afflictions such as influenza and the common cold, mononucleosis, and vaginal yeast infections. There are the rather more serious, but still common, sexually transmitted microbes that cause diseases such as herpes, gonorrhea, and syphilis. There are the persistently gloomy Third World burdens of malaria, cholera, schistosomiasis, tuberculosis, river blindness, sleeping sickness, and a number of others. Of such global pestilences, only smallpox has truly been eradicated. And polio, once nearly as feared in the United States as AIDS is today, has now essentially been eradicated in the Western Hemisphere. There are the other diseases—now nearly forgotten in the industrialized world—of measles, mumps, scarlet fever, whooping cough, diphtheria, and tetanus. And in recent years, the once-forgotten disease of tuberculosis has risen again to threaten many.

Along with this sad catalogue, other invaders seeking to inhabit the human body are responsible for a number of maladies that biomedical researchers are only just beginning to appreciate. There are several viruses that cause cancer, not the least of which is a sexually transmitted entity called papillomavirus that causes tumors ranging from benign warts to potentially lethal cancer of the uterine cervix. It is also beginning to emerge that one cause of mental retardation is a microbe called cytomegalovirus, a virus that inhabits an estimated 80 percent of all Americans but is somehow kept under control in most.

In every moment of every day, every person's body is a battleground where microscopic aggressors constantly seek their biological destiny to multiply and be fruitful. Most of the time, of course, most people remain healthy because their bodies are able to repel the invaders or, at least, to keep them under control. The defenders are the cells of the immune system, the part of the body that, except for the brain, is probably the most complex and least understood. It is not a single organ like the liver or the brain. It is a diffuse system made up of many discrete structures, including the bone marrow, the spleen, the thymus gland, and the lymph nodes, and of many different kinds of free-roaming, independent blood cells called lymphocytes. These are small, round, featureless "white blood cells." Cells of the immune system are spawned in the bone marrow and some receive further rearing in the thymus, but from there they fan out to inhabit every part of the body except the brain.

The immune system as a whole is fabulously complex, a Byzantine array of many cell types with varying powers, each responding to different stimuli but often unable to function except in alliances with various other cell types within the immune system. Most attempts to explain the immune system in the popular press—a favorite metaphor is the military with its layers of command and various types of hardware—fall short. Still, the fundamentals are actually not so hard to grasp if they are considered one part at a time.

The body's first line of defense is its skin and the membranes lining such exposed surfaces as those in the nose and throat. Few microbes can penetrate intact skin, and because its outermost layer is dead anyway, its cells are not vulnerable. Tiny glands in the skin—sweat glands, tear glands, and the sebaceous glands—all secrete antimicrobial chemicals. So do the cells that produce saliva and mucus in the nose, mouth, and throat. In ways that are not fully understood, these cells secrete natural antibiotics that kill bacteria or, at least, block their reproduction.

Within the lungs a more active defense is provided by macrophages, those "big eater" cells that roam the body like hungry amoebas. All kinds of noxious particles, alive and dead, are inhaled into the lungs with every breath. Many stick to the wet surfaces of the tiny air sacs within the spongelike lung. Lurking on the same wet surfaces, however, are macrophages of a type called dust cells, which eat almost any debris they encounter. Enzymes from the dust cell's lysosome digest most of the bacteria and other organic materials, but if the voracious cell swallows something indigestible, it simply stores it in a large vesicle, or vacuole—a kind of internal waste basket. Eventually, when the dust cell gets full and can choke down no more, it seems to know its career is finished. The cell crawls up through the airways, out of the lung and into the windpipe, hauling its indigestible cargo, until the windpipe joins the gullet, which leads down to the stomach. The dust cell is swallowed and digested. Even stomach acid can be considered a defense mechanism because it kills not just dust cells but most swallowed bacteria.

The respiratory system also has a mechanism to rid itself of things that dust cells do not carry out, such as inhaled particles and excess mucus. Lining the windpipe and connecting airways are cells with hairlike cilia that reach into the passageway. The cilia beat in coordinated waves, like the ones that carry the egg in the fallopian tube, moving the unwanted substances up to the gullet and, like the stuffed dust cells, into the gullet.

Though not often considered formal parts of the immune system, all these processes are obviously of major importance in protecting the living cells inside the body.

But it is hardly rare that a microorganism slips through these defenses. Consider a small, simple wound that just breaches the skin, a knife cut or nail puncture. Say there were some bacteria on the knife or the nail and they were inadvertently carried past the skin and inside the body. The immune system's first active defensive reaction is to create an inflammation—the hot redness and swelling that so commonly follow many kinds of injury. Inflammation is probably life's most primitive form of active defense, the first to have evolved. In primitive invertebrates, for example, inflammation is the only defense mechanism.

As soon as the normally tranquil life of the tissue just under the skin is disrupted by the cut, nearby cells initiate the inflammatory response by releasing various substances they have made and stored in vesicles. In many tissues there are special cells, called mast cells, one of whose jobs is simply to wait for trouble. When it comes, mast cells release various substances such as one called histamine. Never more than a few cell thicknesses away are small vessels of the circulatory system. They respond to histamine in two ways. Their cells become flatter than normal, like dough rolled thinner. As a result the blood vessels dilate, becoming larger in diameter so that more blood can flow into the region. Also the cells pull away from one another slightly, releasing some of the junctions that sealed the blood vessel just enough so that the fluid part of the blood, known as plasma, can leak out. As the plasma seeps out, it carries various white blood cells—free-roaming cells of the immune system—with it. As more and more plasma infiltrates the damaged tissue, it causes swelling. When the fluid is distributed generally throughout the extracellular spaces, the result is called edema.

The swelling has no known therapeutic value. It is a byproduct of a phenomenon that does have great therapeutic value—the improved blood flow that delivers immune system cells. The first to arrive are cells that eat other cells, related to macrophages. Their job is simply to consume any microbial invaders and, while they are at it, also gobble up any dead or damaged cells. The first eating cells (the technical term is phagocyte, which is Greek for "eating cell") to arrive is the neutrophil (the name refers not to any function but to its affinity for stains used to see it under the microscope). Macrophages usually show up later.

Within an hour after inflammation begins, circulating neutrophils detect a chemical signal that capillary cells install on their surfaces when they become dilated. The neutrophils stick to the capillary cells and send out tentacles that search for openings between the cells. When a neutrophil finds a hole, it pushes a tentacle into the space and the entire cell squeezes through, rather like sand flowing through the bottleneck in an hourglass. In this way huge numbers of neutrophils make their way out of the bloodstream and into the inflamed tissue.

An hour or so later a second type of white blood cell follows. These are another form of phagocyte called monocytes. They, too, slip out of the capillaries and into the inflamed tissue. Once in the tissue, however, monocytes undergo a developmental transformation. Like Clark Kent changing into Superman, they transform themselves into macrophages. Like all other white blood cells, monocytes/macrophages are produced in the bone marrow and, when mature, migrate out into the bloodstream to become one of the most powerful and widespread agents of the immune system. Blood is teeming with macrophages. There are macrophages crawling about in virtually all body tissues. Macrophages are also abundant in the various epithelial tissues lining body surfaces that come in contact with the outside world—the nose, throat, and digestive tract, for example. Macrophages already residing in the wounded tissue also become activated.

By the time phagocytes—be they neutrophils or monocytes/macrophages—enter the inflamed area in any number, the bacteria there, which may double their numbers every twenty minutes, have been undergoing a population explosion. In all infections, there is a race between the rate at which bacteria multiply and the rate at which cells of the immune system arrive in sufficient numbers to take on the invaders and their progeny. In most cases where a new form of microbe attacks—one that the immune system has not encountered previously—the invaders take an early lead and then are wiped out as the immune system gathers and activates its forces. (Repeat infections by the same microbe are immediately met by antibodies, an advanced component of the immune system which knocks out an infection much faster and which we will get to later. For now, however, we are considering the more primitive immune process of inflammation that occurs with any new infection.)

When a neutrophil or a macrophage bumps into a bacterium, it recognizes the microbe as alien. The phagocyte simply engulfs the bacterium, swallowing it into a bubble of membrane, a vacuole. Inside the phagocyte the vacuole then fuses with a lysosome, condemning the hapless microbe to death by digestion. Phagocytes do not dine quietly. While eating, they exude several proteins that stimulate other phagocytes into a feeding frenzy. Some of the exuded substances stimulate the inflammatory response, further dilating blood vessels, attracting more neutrophils and monocytes, which, of course, become hungry macrophages.

Some exuded substances help regulate blood clotting so that blood is kept flowing where needed but the body is not allowed to bleed to death. And the activated phagocytes also release substances that kill bacteria directly—enzymes that break down bacterial molecules and, amazingly, hydrogen peroxide, which essentially bleaches bacteria to death, and a form of nitric oxide, not to be confused with nitrous oxide, which is better known as laughing gas. The microbial corpses are eventually cleared away by macrophages. Like most forms of chemical warfare, these released substances are rather indiscriminate. They damage many friendly human cells as well as enemy cells. Thus, a tiny cut can become a worse wound, even an abcess, before it gets better.

There is still another defense mechanism that comes into play in an inflammatory reaction to infection. Circulating in the blood is a family of at least twenty different kinds of protein molecules, collectively called complement. Complement molecules are usually in an inactive state. They drift passively in the blood, doing nothing, until a certain trigger molecule shows up. In this case, the trigger is a type of carbohydrate carried on the surfaces of the bacteria. When one of the complement proteins bumps into one of the carbohydrates, the complement instantly changes shape and stays changed. It is now activated. When it bumps into a certain other complement protein, a different one, it too is activated. Number Two then activates Number Three. And so on in a widening chain reaction, much like the cascade of reactions that makes blood clot. The bacterial carbohydrate triggers an elaborate cascade of complement-activating reactions. An arsenal of loose proteins is quickly armed for battle.

Some activated complements signal the capillaries to dilate still further. Some stimulate mast cells to release more histamine, which adds still more to the dilating effect. (It is the increased blood flow to an inflamed area that causes it to feel hot.) Some complements serve as chemical attractants, enticing more phagocytes to the inflamed area. Three types of complement molecule link into complexes that attack the bacteria directly. The three-part structure forms a kind of tube that punches itself into the bacterial cell membrane and stays there, creating a channel through which the bacterium literally leaks to death. Yet another complement protein simply attaches itself to bacterial surfaces, marking them for faster consumption by phagocytes. In technical language, incidentally, biologists call this process opsonization; that's from the Greek for "to prepare for dinner." Neutrophils and macrophages carry receptors for this particular complement protein.

The entire inflammatory process is an elaborate network of interacting cell types and loose molecules of many kinds. Parts of the network contain powerful feedback loops in which a small initial response can mushroom quickly into a massive one. For example, a little inflammation can lead to the production of substances, such as histamine, that rapidly increase the amount of inflammation, which releases still more histamine. And so on. Normally the interactions are delicately balanced, and a slight imbalance can lead to runaway chain reactions that turn a minor wound into a major episode. This is why one person sustaining a slight injury will have little or no reaction and another, with only the slightest difference in body chemistry, will swell up for days. Allergic reactions are a form of this phenomenon, but they usually involve antibodies, which are parts of the more advanced immune system that we'll get to next. At that point, we'll return to the subject of allergy and other afflictions in which the immune system goes out of control in a major way.

Eventually, the various agents of the inflammatory reaction vanquish the bacterial invaders, the inflammation subsides, and the damaged tissues heal.

Once the immune system has encountered a new type of microbe with this generalized reaction, which is relatively slow (and some infectious agents can cause serious disease before they are wiped out), it prepares to respond more quickly the next time the same microbe attempts another assault. The immune system does so through the more advanced mechanism of specific immunity that evolved to accompany the primitive, and generalized, inflammatory reaction. Unlike inflammation, which is often painful, specific immunity is usually so effective that it functions without a person ever being aware of the battles within.

Once again, macrophages play a key role. The voracious cells that gobble up the bacteria, viruses, or whatever, in a generalized immune reaction don't just eat and run. They help prime the specific immune system to defend the body should the same kind of bacteria ever come back again. When the lysosomal enzymes digest the bacteria eaten by the macrophage, they don't break everything down to molecular smithereens. The proteins that were on the surface of the bacteria, for example, are only partly digested. Sizable fragments of them are transported out of the digestive vacuole and installed in the macrophage's outer membrane. The macrophage then "presents" these fragments to several other lymphocytes in a form that induces the other cells to join the attack on any remaining bacteria bearing the same proteins on their surfaces. The macrophage's action is a bit like the county sheriff giving his bloodhounds a sniff of the fugitive's clothing, essentially telling the other cells how to recognize the enemy. Immunologists call these fragments antigens, a term that refers to any molecule that will stimulate a response by the immune system. Of course, the immune system is a little more complex. It has several kinds of sheriffs—other types of cells that also present traces of the perpetrator, or antigens, to the bloodhounds—and all the cell types must work in teams, each stimulating the other into action.

In the steps that follow, the central player is a lymphocyte, often called the "helper T cell," which, as its name implies, stimulates the other kinds of cells to action. Unfortunately, different naming systems have been used in immunology, and this cell is sometimes called a T-4 cell or a CD-4 cell. Incidentally, the helper T cell is the chief one targeted by the AIDS virus. HIV, the human immunodeficiency virus, invades the helper T cell and splices its genes in among those of the host cell. Then the viral genes command the cell to manufacture hundreds of new viruses, driving it to a prodigious rate of activity. Eventually, the new viruses burst out and the helper T cell is killed. Because its helping role is so crucial to the working of the specific immune system, killing off a large number of a person's helper T cells can seriously weaken his or her immune system. To make matters worse, HIV also invades and kills macrophages. This is the chief action of the virus's attack on the brain, leading to AIDS dementia.

Another key player in the specific immune system is the B cell. ("T" and "B," incidentally, refer to organs that played roles in the cell's development. All the cells of the immune system arise in the bone marrow. Those that go directly on duty to make antibodies are called B cells. The B comes from "bursa of Fabricius," the organ in birds that controls the development of antibody-producing cells. Those that go first to the thymus, an organ just above the heart, and mature there are called T, for thymus-derived cells.)

Antibodies made by B cells are protein molecules of a particular shape that will bind only to molecules having a precisely complementary shape. Each antibody is shaped like a Y. Each arm has a special shape that acts as a receptor for some specific molecule. Antibodies may remain bound to the B cell, attached by the stem of the Y, or they may be released to drift free in the bloodstream. Any molecule to which an antibody will bind— be it a protein, carbohydrate, or nucleic acid—is, by definition, called an antigen. Although all B cells are fundamentally the same, they differ spectacularly in the types of antibodies on their surfaces. Each B cell is studded with only one type of antibody, but there are hundreds of thousands, and possibly millions, of different shapes of antibodies. In the blood and in the body's lymphoid organs, such as the spleen and the lymph nodes through which blood or plasma circulate, there are countless B cells, each capable—to continue the bloodhound analogy—of sniffing out only one among thousands of different possible fugitives—the one antigen its antibodies are shaped to fit.

When the right B cell happens to bump into a macrophage presenting an antigen that fits its antibody, the B cell is instantly activated for battle. The antibody does not attack the presenting lymphocyte, however, because it also carries another molecule on its surface that the B cell recognizes as the badge of a friendly cell. The activated B cell is ready for battle, but it is just one cell, and one cell is rarely a match for a bacterial invasion. This is where the helper T cell comes in, and what happens next is one of the most complex processes in biology, one whose exact mechanisms are still being worked out.

Helper T cells also carry their own version of antibodies, which are actually receptors that stay bound to the T cell's membrane. When the receptor binds briefly to the antigen presented by the macrophage or other lymphocyte, the T cell becomes activated. While linked to the helper T cell, the presenting cell secretes a chemical that the helper T cell takes up. It is a stimulant for T cells to divide repeatedly. The proliferating T cells then secrete another kind of stimulant that acts only on activated B cells. It makes them start dividing.

In this way, the immune system selects only those B cells bearing the right antibody and stimulates only them to proliferate, steadily raising the population of B cells capable of dealing with the particular bacterial invasion at hand. Many of the multiplying B cells, which live in the lymph nodes, spleen, and other locales within the body, shed their antibodies to drift with the bloodstream. Passively, the antibodies fan out and track down the bacteria almost anyplace they might have spread in the body. As antibodies encounter bacteria bearing copies of the antigen, the antibodies lock on to their targets with two deadly effects.

First, any bacterium coated with antibodies looks especially appetizing to a phagocyte, either a neutrophil or a macrophage. The antibody-marked bacterium is likely to get eaten very quickly. And second, B cells can also make antibodies that bind not to the whole bacterium but to the toxic proteins the bacteria give off and which cause much of the disease attributed to the bacteria. When an antibody binds to a toxin molecule, the toxin can no longer react with the body to cause illness.

If the infection is caused by a virus, antibodies can bind directly to the outer surface of the virus. Ordinarily this is a bigger threat to a virus than it is to a bacterium because of a fundamental difference between the two. Bacteria are full-fledged organisms, although they have only one cell. They feed, grow, and multiply under their own power. Viruses are not even a single cell. They consist mainly of a short piece of DNA or RNA, which encodes a few genes—often fewer than a dozen—enclosed in a protein shell. Viruses cannot perform any metabolic processes and are essentially inert. They cannot multiply by themselves because they lack the machinery to read their genes and follow the instructions. For this reason, viruses are not really alive in the usual sense. The only way a virus can multiply is to invade a cell and hijack its metabolic machinery. Viruses get into cells because parts of their coat proteins masquerade as proteins that some cell type normally takes in via a receptor.

In the case of HIV, for example, the virus's coat protein happens to include exactly the right shapes to fit receptors on the helper T cell. When the virus binds to the receptors, the cell "thinks" it has a hold on something it wants and it pulls the virus inside the cell. Since helper T cells are the main ones carrying those particular receptors, they are the natural target of the AIDS virus.

All viruses get into cells in essentially the same way. So, when antibodies bind to proteins of the virus's shell, they block the virus from invading a cell. They do so simply because the antibodies are physically in the way. The virus protein can't make contact with the cell's receptor. One of the sad discoveries of the AIDS epidemic is that antibodies to HIV, even though they may block access to the receptor, do not stop the virus from invading cells. HIV has other means of spreading directly from cell to cell without ever venturing out into the open bloodstream where the antibodies lurk.

While the early phase of the immune system's defense is underway, each activated B cell also grows fatter as its endoplasmic reticulum (ER), the first organelle that processes newly made amino acid chains, becomes larger and larger. The extra ER makes it possible for the cell to step up its rate of manufacture until it can turn out several thousand antibody molecules per second. Activated B cells live only a day or two in this accelerated mode. Then they die. This is one way the immune system keeps its defense proportionate to the attack. As long as there are infectious microbes at large, the whole process of macrophage eating, helper T-cell activation, and B-cell activation continues, keeping up the production of antibodies. But as the enemy dwindles, so does the attack.

Once a virus is inside a cell, however, the immune system cannot "see" it. The cells and antibodies of the immune system simply have no way of getting inside an infected cell to get at the virus. As it happens, however, cells that have made the mistake of taking in a virus cannot keep their secret. They soon reveal themselves and certain specialized cells of the immune system—called cytotoxic T cells—quickly find the infected cells and kill them, dooming the virus inside.

What happens is that as the virus's genes are commanding the hapless cell to manufacture new virus coat proteins (in preparation for assembing whole new viruses), the coat proteins are first installed in the cell's outer membrane. This the cytotoxic T cell easily spots because it is equipped with a receptor—shaped like an antibody to the protein—that binds to the viral protein. The T cell immediately releases substances that act on the target cell, inducing it to activate its built-in suicide program. The human cell must die, but that death dooms the virus that invaded it and prevents the cell from making lots of new viruses.

In addition to inducing suicide, the immune system may also protect the body against long-term infections by certain microbes within cells by committing cellular homicide. A special kind of helper T cell, upon recognizing that cells within various tissues are inhabited by foreign agents, starts producing chemicals that stimulate monocytes to undergo the Clark Kent transformation mentioned earlier, becoming voracious macrophages that eat the infected cells.

Why don't these processes wipe out cells inhabited by the AIDS virus? Because one other factor is needed. As with the relatively few B cells that happened already to be equipped with antibodies to fit a microbial type that is invading the body for the first time, there are relatively few macrophages and cytotoxic T cells with the right receptors to fit, and thus recognize, the foreign proteins on the infected cells. It takes time for them to proliferate. The stimulant to drive this proliferation is a hormonelike molecule called interleukin-2, or IL-2, made in large quantities by the central coordinator of the immune system, the helper T cell. IL-2 is one of more than forty "immune hormones" called cytokines, which regulate the growth and function of the various cells of the immune system, including the helper T cells themselves.

And, of course, this is the very cell attacked and killed by the AIDS virus. People with AIDS often have dwindling numbers of helper T cells and, therefore, dwindling amounts of IL-2 and other cytokines needed to make cytotoxic T cells proliferate into quantities that can do any good. Or to make large numbers of monocytes change into macrophages. It is facts like this that make AIDS researchers regard HIV as the most diabolical of viruses. It destroys the very heart of the body's mechanisms for defending against viruses and, of course, all other kinds of infections.

A similar form of cell-killing is performed by yet another kind of cell, one that goes by perhaps the most explicit name in immunology—the natural killer cell. Like the cytotoxic T cell, it emits substances that trigger the target cell's suicide program. But the natural killer, or NK, cell's targets are not primarily those infected by microbes. It goes after defective native human cells, including those that have turned cancerous. These are cells that have undergone several genetic mutations (that's what made them malignant in the first place) and which are making abnormal proteins that are carried on their surfaces. Many researchers think that if it were not for NK cells, people would be getting cancer all the time. (You will find more about cancer in the next chapter.)

Once the infection is defeated, the body becomes a slightly different place. It is now immune to a new attack by the same kind of invader. This is because not all the activated B cells die off. While most grew into the fat, antibody-making machines with the short lifetimes, a few did not. They became what immunologists call memory cells. These cells may live for many years, providing the body with a much larger population of cells specifically equipped to deal with one particular antigen carrier should it ever show up again. Before the first invasion of this variety of microbe, the body had very few B cells carrying antibodies with the right shape. It took time for one of those cells to encounter a macrophage or other immune cell presenting the antigen. Now, however, the body has lots of the right kind of cells waiting, preserving the memory of the kind of antibodies and receptors needed to find and attack that particular strain of pathogen. Chances are the memory cells will mount a successful defense before the invader can multiply into numbers that threaten health.

In all of the preceding explanation, one crucial detail has been glossed over. For many decades it was one of the greatest puzzles of immunology. Experiments on animals show that no matter what foreign substance one injects into an animal, its immune system will make antibodies precisely shaped to fit—lock and key again—those proteins. How can this be? There are millions of kinds of proteins, carbohydrates, and other antigenic substances in the natural world, and it seems inconceivable that every immune system comes equipped with the complete knowledge of the shape of every one of them. Experiments even show that if scientists create an artificial protein, one with a shape not known in nature, and administer it to mice as an antigen, the immune systems will come up with exactly the right antibodies in a day or so—as fast as for any natural protein.

Because each antibody is itself a protein molecule and because we know that proteins are encoded by genes, it seems there would have to be a gene encoding the shape of every conceivable antibody. It is estimated that the human body can produce somewhere between one million and a hundred million different kinds of antibodies. But there are only between 50,000 and 100,000 genes in the human genome. So how does the immune system do it?

The answer began to emerge only in the mid-1970s. It has turned out that there is not a gene for every antibody. Instead the genome contains a few hundred different antibody gene segments that can be mixed and matched and linked in many different combinations. When B cells are developing in the bone marrow, their precursor cells literally shuffle the gene segments in their kit and settle on random combinations. The resulting cell then makes its antibody, carries it around on its surface, and waits to see if it works. If the antibody key never finds an antigen lock that fits, the cell that made the antibody leads a very dull life. It never becomes activated and therefore can never be stimulated to proliferate.

But if the randomly combined gene segments produce an antibody that does fit an antigen at least roughly—close enough for the cell to become activated—the cell will multiply at a modest rate. Then, with each round of cell division, the antibody genes in the daughter cells undergo minor changes within their DNA. This time, DNA segments are not shuffled; that would create radical changes in the antibody's structure. Instead, the DNA is mutated slightly at specific sites, creating numerous variations on the existing structural theme. As a result, the next generation of B cells is likely to contain one or more cells whose antibodies are a closer fit than any in the previous generation. If so, those cells are selectively, or preferentially, activated and stimulated more powerfully to proliferate. And again the genes dictating the antibody structure are mutated slightly. And if a still closer fit is achieved between antibody and antigen, the cell bearing that antibody is stimulated to proliferate even more abundantly.

What is happening is a form of natural selection—the phenomenon that makes evolution occur. In nature, the species best adapted to its environment, reproduces more prolifically than can the species not so well suited. The better-adapted species comes to dominate the environment, and its progeny, bearing random mutations of the parents' genes, may come up with a specialization for the environment that makes it even more successful. In this way a series of fine-tuning steps produces a species consummately adapted for its ecological niche.

It is the same with the immune system. The better an antibody fits an antigen, the more the cell that made it will divide. Evolution in nature takes thousands or millions of years to produce a significant change. Evolution of B cells in the immune system usually produces a perfect fit of antibody to antigen in a matter of days. Still, those days while immune cell evolution is going on are the days when you're getting worse at the beginning of a cold or the flu or some more serious microbial attack. Then, as the immune system perfects its antibodies, the tide of the battle gradually turns. Before long, the immune system is churning out defensive forces faster than the offensive forces can multiply. A few days later, the last of the invaders is killed off and the body is left with a rich supply of cells perfectly tuned to repel any new attack by the same virus or bacterial species. Among these are the memory cells, the special immune cells that preserve the precise genetic sequence needed to make the right antibody and simply wait for another attack.

This is the process that vaccines are designed to stimulate. A vaccine is nothing more than a method of tricking the immune system into thinking a real infection is underway. For example, a vaccine may contain dead bacteria of the species that causes a given disease. The bacteria can't reproduce or cause any illness but the microbial corpses still carry their characteristic surface antigens. The immune system, which can't tell a live bacterium from a dead one, sees the antigens, recognizes them as alien, and mounts an attack. Macrophages gobble up the dead bacteria and dutifully present some of the bacterial proteins on their surfaces, thus alerting other agents of the immune system to make the right antibodies and specific T cells (ones carrying appropriately shaped receptors) and head into battle. The memory cells created in this process never realize there is no battle. They obligingly take up their job of waiting for the next attack. Sometimes a single dose of such a vaccine leads to only a few memory cells or to memory cells that are not as fully evolved as possible to suit the microbe. Booster doses of vaccine are often used to continue the process where the last dose left off.

Some vaccines use dead bacteria or "dead" viruses. (Since a virus is not really alive in the sense that a cell is, it can't be dead either. Vaccine makers chemically treat viruses so that they cannot invade cells and call them "killed virus" vaccines.) Some vaccines use live bacteria or "live" viruses, which are microbes that have been modified to render them unable to cause disease but still able to multiply in the body. Vaccines made from live agents are often more effective because as the microbe multiplies, it makes its own booster doses.

One of the newest forms of vaccine contains no microbe at all, just some of its antigens. This approach is being tried for microbes that are too dangerous to risk injecting alive or dead into a person. For example, the process that kills or cripples the microbe can sometimes fail, and somebody expecting to be protected could be given the disease instead. Exactly this mishap occurred with the Salk polio vaccine in the early 1950s. Some 260 vaccinated children were inadvertently given polio, and 11 died before the mistake was discovered and manufacturers found a more reliable way of inactivating the poliovirus. Today many vaccine researchers are simply unwilling to risk injecting people with the AIDS virus in any form. Instead, they are trying to develop a vaccine made only from some of the proteins of the AIDS virus.

The oldest form of vaccine—the form from which the very word vaccine comes—involves a still different approach. In the late eighteenth century, the English physician Edward Jenner developed the world's first fully effective and completely safe vaccine against smallpox, a disease that until recently was one of the world's scourges. Jenner had learned from farm folk that people who got a very mild disease called cowpox, because they worked with infected cattle, never got smallpox. The inadvertently immunized were usually people who got some of the pus from the cows' sores into scratches on their own body. Jenner figured that if people were deliberately given cowpox, they would also be immune. He took some "matter" from the cowpox sores of milkmaids and rubbed it into fresh scratches on his patients. Jenner didn't know anything about the immune system, but it is clear now that the scratches were wounds in which the cells of the immune system could encounter the cowpox virus.

Jenner's preparation was called "vaccine" from the Latin *vacca,* for "cow." Nowadays, the cowpox virus is called vaccinia. Jenner's vaccine worked because of a lucky circumstance. Vaccinia virus and smallpox virus have surface antigens that are virtually identical. Antibodies primed to bind to a vaccinia antigen will bind about as readily to a smallpox antigen. Since the vaccinia virus was not "killed" or even weakened, it multiplied in the body, providing the sustained challenge that eventually makes the immune response highly effective.

In the 1970s smallpox was eradicated, the first disease ever relegated to this status. A relentless worldwide vaccination campaign by the World Health Organization tracked down every person remaining who had been exposed to the smallpox virus—the last cases were in the Ethiopian bush—and vaccinated them. Since then not one person has gotten small-pox, except for one or two people accidentally exposed in laboratories doing research on the virus.

But the vaccinia virus has not been retired. With two centuries of experience attesting to its safety, medical researchers are now experiment-ing with ways of genetically modifying the old war-horse so that it carries surface proteins copied from those of other microbes. That way the virus will invade the body and cause a mild and usually unnoticed infection while provoking the immune system to make antibodies and memory cells tailored to fight all its antigens, including the spliced-in one from some other microbe.

Some decades ago, when surgeons began trying to cure people by trans-planting replacement organs from other people's bodies, medical science began to grapple with one of the most puzzling aspects of the immune system. Almost invariably, the patient's immune system would attack the introduced organ and kill it. How could it do this? How could mere cells tell the difference between the tissues of their own body and those that formed in another body?

Answering such questions has been one of immunology's greatest chal-lenges, and the answers themselves are extraordinarily complex. But the basic principle is simple. Every cell in the body carries a cluster of special marker proteins on its outer surface. This is the "badge" referred to earlier that identifies friendly cells. The genes for these proteins come in so many different versions that the combination displayed in each person is essentially unique (except for genetically identical siblings, who carry identical markers). The cells of the immune system recognize the marker proteins of their own body (they have receptors shaped to fit only these identifying badges) and will attack only those cells that carry markers of any different configuration. These marker proteins are called the "major histocompatibility complex," or MHC. (The ones on human cells are also called human leukocyte-associated antigens, or HLA antigens.)

Though the principle seems simple enough, the discovery of this sys-

tem raised a profound question. MHC proteins are found only on vertebrate cells and not on the cells of any of the microbes or invertebrate
parasites against which the immune system is supposed to defend the
body. Why would an immune system that evolved long before the existence of organ transplantation have developed the ability to detect and
attack tissues from another vertebrate? Vertebrates don't invade other
vertebrates, so there would never have been a need for such a defense
mechanism.

The answer came when immunologists discovered the exact mechanism by which T cells learn to recognize foreign antigens from microbes
and invertebrates. As mentioned at the beginning of this chapter, the
invaders are first recognized only when fragments of their proteins are
"presented" on the surfaces of macrophages that have eaten them. Presen-
tation can also occur on other cells in which viruses are reproducing when
copies of newly made coat proteins appear on the cell's surface.

T cells have receptors shaped to fit only the MHC proteins of their own
cellular republic—the passports, if you will, of their fellow citizens. They
go around constantly checking the passports of other cells. If the MHC
passports are in order, the cell is allowed to go on its way. But MHC
proteins are built to perform a second function; they stick to foreign
protein fragments that appear in the cell membrane. Once this happens,
the shape of the overall MHC-fragment complex is no longer recognized
by the T cell. Its receptor recognizes the MHC part but finds an attached
structure that shouldn't be there. The T cell reacts quickly, sending out
the chemical signals that recruit other immune system cells and start the
process of killing the cell bearing the falsified passport.

The same mechanism, it turns out, comes into play with a transplanted
organ. Because its MHC is wrong to start with, that's enough to set off the
T cell and launch the immune system's attack. Doctors can get partway
around this by seeking organ donors with "matching" tissue types. These
are donors whose cells bear marker proteins that are more like those of
the intended recipient. Experiments have shown that the more similar are
donor and recipient in these proteins, the less likely the recipient's body is
to reject the donated organ. An added way of overcoming transplant
"rejection," as the phenomenon is called, is to weaken the patient's immune system with drugs.

Because of their ability to distinguish native cells from foreigners, most
immune system cells are well behaved most of the time. They politely
refrain from attacking the proteins and cells of which their own body is
made. Somehow, however, the immune system can lose the ability to tell
self from nonself. The policemen of the republic of cells can suddenly turn
disloyal, rampage out of control, and attack their own citizens. Immune
system cells may, for example, shuffle and mutate their genes into combinations that prime the B and T cells to recognize perfectly proper and

normal proteins of their own body. The attacks can be fatal to one's own cells, killing them as easily as they do bacteria.

The result is what are called autoimmune diseases. Renegade immune systems, unfortunately, are not rare. They cause several relatively common diseases. These include multiple sclerosis (the immune system attacks the special cells that form an insulating layer around nerves, causing nerves to "short out" just like wires that have lost their insulation), rheumatoid arthritis (the immune system attacks the joints, breaking down tissue and causing painful inflammation and swelling), myasthenia gravis (the immune system attacks receptors on muscle cells that receive chemical signals from the nervous system, leaving the muscles paralyzed because they no longer get the signal to move), and type I diabetes (the immune system attacks the special cells within the pancreas whose job is to make insulin). In all of these diseases and many others less well known, the immune system has lost its ability to distinguish a foreign antigen from one that belongs in the same body. The immune cells attack the native cells as vigorously as if they were mortal enemies of the body. Some medical researchers suspect that many more diseases that are not understood will turn out to be the result of immune systems gone bad.

Unruly immune systems are not always so deadly. As countless millions know, they can merely make you miserable. Allergies are the result of the immune system reacting to foreign antigens, called allergens in this case, carried as part of pollen, dust, food constituents, cat dander, and—unfortunately for a few people—practically all other fragments of the natural world. In most allergies, the immune system produces some conventional antibodies, but the misery comes from the effects of "immunoglobulin E," or IgE, a special type of antibody produced by a subclass of B cells. Its normal job is to defend the body against invasion by parasites, organisms larger than bacteria or viruses. This antibody, however, does not attack the foreign antigen carrier directly. Instead it attaches itself to mast cells, the immune system components that play a role in inflammation. This triggers them to release massive amounts of histamine and other substances that dilate blood vessels and stimulate the chain reaction of events that causes inflammation.

While this response is often the best the immune system can do against large-bodied parasites, it does no good when triggered by otherwise harmless allergens. If the allergen is inhaled, the inflammatory response happens mainly in the nose and throat, which begin swelling and secreting mucus as if they were infected by a cold virus. Instead of a real fever, one has hayfever, complete with the sneezing, congestion, and difficult breathing.

The reason some people are allergic seems to be because their immune systems are not properly regulated. They make too strong a response to environmental antigens and produce too many of the cells that make IgE, the special antibody that tells mast cells to make histamine. People who have very large populations of these cells in their blood can have reactions so severe that they spread from the original site, say the nose and throat, and cause the entire circulatory system to dilate and leak fluid. This so-called anaphylactic shock can cause the circulatory system to fail and kill the person. Every year, for example, a few hundred people die this way from a single bee sting.

The immune system isn't perfect. But it is very close to it. Inconceivably large numbers and kinds of microorganisms try to inhabit the human body and feed upon its organically rich contents. But as long as the immune system is healthy, they don't succeed—at least, not for long. The reason that mold grows on bread or that meat rots in the garbage can is because neither has an immune system. Without ours, quite simply, we would all long ago have rotted away.

Cancer is one of life's most fascinating phenomena. To be sure, it is a fearsome threat, but as cell and molecular biologists learn more about how the disease starts and develops, they are learning that cancer has an extraordinary natural history, one that is shedding light on the most fundamental workings of life itself, even as it moves medicine closer to the conquest—if such a thing is possible—of cancer.

The demand for a cancer cure has surely been one of the strongest drivers of public and private investment in biomedical research. Announcements of "new hope" and near-cures have long peppered news reports and repeatedly been toned down or dropped altogether. And every few years, hope has been tempered when further discoveries revealed cancer to be even more complex and resistant than was previously thought.

Now, as we approach the twenty-first century, a new wave of optimism has emerged—this time not from the clinical side of science but from a remarkable convergence of basic research in several once-disparate specialties far from the laboratories or bedsides of conventional cancer research. For example, the discoveries of people studying cell division for its own sake, even in yeast cells, are merging with the theories of evolutionary biologists. And the findings of those who probe the mechanics of DNA are merging with the remarkable new understanding of cellular suicide, or programmed cell death. From all these fields, plus what mainstream cancer research has learned, is emerging a unified theory of cancer that may, just may, provide the basis for the long-elusive dramatic gains in treating and curing cancer.

Far from being simply the result of rapid, disorderly cell proliferation, cancer is revealing itself to be a highly logical, coordinated process in which certain cells—as if they were plotting a revolution within the republic of cells—acquire an array of specialized abilities that allows them to pursue their renegade cause. The cancer cells that arise in a person are the product of a kind of high-speed evolution in which random mutations arise at a furious pace in the genes of individual cells. Thereupon, natural selection within the body favors the survival of those cells most able to proliferate without control and in places where they shouldn't. This internal evolution also favors those cells that happen to acquire the ability to leave their native tissues and strike off into new habitats, other organs of the body. The phenomenon is essentially identical to that of Darwinian natural selection. The "fittest" of the cells to evolve eventually becomes the deadliest because its offspring have inherited a variety of altered genes that confer a suite of powers that give them a kind of functional superiority over the more cooperatively inclined powers of the republic's normal cells.

New research findings also indicate that the sequence of events that gives rise to cancer involves several of life's most fundamental processes, subverting them from their original benign, even essential, roles in the cell to the most dangerous of ends.

For example, there is now evidence that some tumor cells proliferate out of control by commandeering a mechanism that cells are "supposed" to use only to repair wounds. In many cases cancer cells hijack the genes and enzymes that exist to spur skin cells and others to heal a wound. In other instances, tumor cells reactivate genes whose main purpose is to operate during embryonic development but which are then supposed to shut down for good. In acquiring—or reactivating—these abilities, many cancer cells revert to the undifferentiated state of a cell in an embryo, throwing off the specializations that restricted their freedom to multiply, their freedom to get up and roam off to new parts of the body, their freedom to settle down in far-flung tissues and put down roots, and their freedom to reproduce, passing on their lawless ways to their many descendants.

The attention given in recent years to "cancer genes"—the so-called oncogenes—has led to the misconception that we all carry around cancer-causing genes in our cells, but this is not so. The genes in question are entirely normal and necessary for life. They are, however, *potential* cancer genes, or proto-oncogenes, as some experts like to call them, because after undergoing certain abnormal changes in their genetic sequence, the modified genes can turn a cell cancerous. The change can be a point mutation within a gene—as simple as substituting one DNA base for another—or it can be a rearrangement within the gene, or it can be the accidental pairing of a gene with a regulatory sequence that drives the normal gene faster than normal.

Whatever the change, it is now clear that one alteration, or mutation, in the code is not enough. Several genes—as few as two in one form of cancer to perhaps ten or twenty in other types—must be changed to transform a well-behaved, specialized cell into a rampaging killer. If the right mutations happen, a cell will surely become cancerous, but those changes come at the end of a long and improbable chain of causation.

Before cancer can start, in fact, a whole series of rare events must occur. Indeed, the right combination of events is so unlikely that many cancers do not arise unless cells grow and divide many times during several decades of exposure to the environmental factors such as cigarette smoke, ultraviolet light, or other cancer-causing agents that start the process. Many people live healthy lives with some of their cells having evolved part of the way toward being cancer cells and then die in old age of other causes before all of the evolutionary steps can be completed.

The likelihood of this sequence of events happening in just the right way looks so overwhelmingly improbable that it seems a wonder anyone gets cancer. But of course millions do. The fact that one out of four Americans will develop cancer affirms that exposure to the gene-altering causes must be widespread and persistent. Still, new discoveries do offer hope. The series of steps, or missteps, that must occur to produce a malignant tumor offers many points at which it may be possible, with further research of course, to intervene and block the end result.

The cancer process starts in many people through contact with cancer-causing substances, or carcinogens, such as the benzopyrene found in tobacco smoke and many other sources. Contrary to popular impression, however, chemical carcinogens are not always harmful in their original form. They arrive in the body innocuous. They are then turned into potential killers by the body itself, by special cells whose job is supposed to be to detoxify poisons that get into the body. These detoxifying cells—found mainly in the liver but also in skin, placenta, and other organs—chemically alter various unwanted molecules into a form that is more easily excreted. Thus, when carcinogens enter the body, all, or nearly all, are properly detoxified and eliminated from the body.

Researchers at the National Cancer Institute have found, however, that people differ genetically in their complement of detoxifying enzymes. Errant enzymes may perform the wrong modification to carcinogen molecules. Instead of rendering them harmless, the enzymes alter the molecule so that it becomes more potent—better at slipping into a cell's nucleus, perhaps, or more avid in its ability to bind to DNA in a way that affects a gene's activity. This modification of the carcinogen, called activation, is the first step toward a cancer-causing mutation.

"We've known for a long time that people differ in their susceptibility to cancer," says Harry Gelboin of the National Cancer Institute, "and it just may be that this relates to relative amounts of the enzymes. If you happen to have the wrong combination, carcinogens may be a lot more dangerous for you than for your neighbor." Gelboin suggests it may someday be possible to test people for their detoxifying enzymes and see who might have the riskier forms.

Although activated carcinogens are able to bind to DNA, most probably do not because cells have scavenger molecules that readily grab on to alien molecules, rendering them harmless. Moreover, most activated carcinogens that elude the scavengers are thought to attach harmlessly to proteins and other cell constituents besides DNA. It is the rare activated carcinogen molecule that makes it into a nucleus and binds to DNA. Once that happens, the risk of cancer goes up dramatically because it can damage a gene, altering its sequence.

DNA damage can also come from ionizing radiation, and it can happen spontaneously, a result of the random banging about of various molecules within the nucleus, including highly reactive oxygen radicals. The result can be to change the chemical nature of a base or to break the link from one to another or even to form bonds between adjacent nucleotides that cannot be broken. Over the course of a lifetime, these spontaneous DNA-damaging events are estimated to happen to every single gene about 10 billion times.

Even then, harmful consequences are unlikely, for cells have yet another line of defense—a built-in DNA repair mechanism. Special enzymes in the nucleus patrol the DNA and can detect abnormalities such as an alien molecule clasping a DNA strand or even a potentially mutating defect in the DNA produced by a hit-and-run carcinogen. The repair enzymes perform surgery on the DNA double helix, cutting out a small segment of the strand containing the problem and replacing the excised portion by assembling a new complementary sequence using the opposite strand as a template. This usually works because the damage is often confined to just one of the two complementary strands.

If DNA repair occurs before the cell undergoes its next division, the cancer process is blocked. But if the repair mechanism is faulty or if the cell divides before repair can take place, the affected part of the genetic code may be copied in its mutated form. The daughter cells will then inherit a mutated gene in a form that no DNA repair mechanism can detect. To the repair mechanism, the changed part of the DNA now looks perfectly normal. From now on all the progeny of the mutated cell will inherit the mutation.

Once cancer researchers realized the role of the DNA repair mechanism, they realized that flaws in the repair process could predispose a person to cancer. There is also clinical evidence of a link. James Cleaver of the University of California at San Francisco discovered the best-known link in people who have a rare skin disease called xeroderma pigmentosum. All the cells of their body have inherited a defect in the DNA repair mechanism that corrects the main form of damage caused by ultraviolet light. Consequently, victims of this disease have an extremely high rate of skin cancers.

Several other diseases marked by strong predispositions to cancer have also been found to involve defective DNA repair mechanisms. Although each is rare, some researchers suspect the prevalence of faulty DNA repair may be high enough to account for as many as 20 percent of all human cancer cases.

A mutation, however, does not automatically mean cancer. Since the alien molecule may bind to any of a cell's genes, most of the mutations that result are likely to have little or no effect. Most of the genome consists of sequences other than genes. Even then, most genes are inactive in any given cell. And, of course, the alteration may simply kill or cripple just the one cell. The mutation that leads to cancer must be of a very specific nature. For one thing, its effect must not damage the cell's vital housekeeping processes. Moreover, the mutation must grant the cell new powers, throwing off the shackles that kept it a loyal and docile subject of the republic of cells. If all these things happen, the long intracellular process that leads to cancer has been initiated. In fact, cancer biologists call the carcinogenic agent that caused that first step the "initiator."

Many experiments in which people have tried to induce cancer in laboratory animals have established that at least one more mutation must take place and, in most cases, several more. There is ample evidence that tumors are made up of the progeny of a single, founder cell that has turned cancerous. So the subsequent mutations must happen in the cell that harbors the first mutation. Since this cell may be only 1 among, say, 100 billion lung cells, the odds of all the previous steps being repeated in the same sequence so that a mutation of a particular other gene happens in the very same cell make the odds of cancer vanishingly small—at least for any individual cell.

"What has to happen," says Stuart Yuspa, another National Cancer Institute researcher, "is some process that improves the odds for progression in that lone cell." In other words, the odds of a cell accumulating enough mutations are increased if the cell attacked by the initiator—or that cell's progeny—also undergoes some change that makes the cell more likely to divide. Yuspa's research on laboratory animals and cultured cells has helped to explain how this happens. And it has helped explain what happens during the long delay between exposure to carcinogens and the appearance of tumors—the famous latency period that means cancer seldom shows up until long after the victim has been exposed to a carcinogen. It typically takes ten to twenty years of cigarette smoking, for example, to produce the first lung tumor. And the incidence of leukemia among residents of Hiroshima and Nagasaki did not begin to appear until five years after the nuclear attacks and peaked another three years later.

Yuspa's experiments involved painting known carcinogens on the shaved skin of mice. Usually there is no obvious change. Even after a considerable length of time, no tumors develop. But if the mouse is also dosed with certain other chemicals, known as promoters (not to be confused with the promoter regions in DNA), tumors will arise in the area treated earlier with the initiating carcinogen. Mice exposed only to the promoters get no tumors.

Scientists interpret such experiments as indicating that the first exposure damaged the DNA and caused a mutation but that its effects could not show up until the cell was stimulated to divide more frequently. Mutation somehow sensitizes the cell to the actions of the promoter. The promoter causes the mutant cell to proliferate more than do neighboring cells lacking the first mutation. As long as the cells are exposed to the promoter, their numbers grow faster than those of other cells, gradually increasing the odds that one of the cellular descendants—still carrying the original mutation—will get a second DNA hit.

"When you do the statistics—the rates of cell proliferation and cell death, the chances that mutations will be at the wrong sites, and so forth—you begin to see why it should take years and years to produce a tumor," Yuspa says. The statistics also suggest why cancers are the most common among cells that already have a high rate of cell cycling. More than 90 percent of cancers in adults arise in just one type of tissue, the epithelial cells that make up skin and the lining of such organs as the gastrointestinal tract, the uterus, the lungs and airways, and the glands. (Cancer of epithelial cells is called carcinoma. Other fairly common forms of cancer include lymphoma [cancer of the lymphoid system, which is part of the immune system]; leukemia [white blood cells]; sarcoma [cells of the connective tissue, which includes muscle and bone]; myeloma [bone marrow cells].) Cancer is extremely rare among cell types that never divide such as adult nerve cells. Brain cancer, incidentally, usually arises not from the neurons but from glial cells, which wrap themselves around neurons and help support them.

Many chemicals in the human environment can act as promoters, each usually exerting its strongest effect on a specific cell type. Phenobarbital, for example, is a strong promoter of liver cells. Cigarette tars contain not only initators but promoters that speed the proliferation of lung cells. Saccharine and cyclamate, the two controversial ingredients of certain artificial sweeteners, are each weak carcinogens but stronger promoters. Certain kinds of dietary fat appear to be indirect promoters, acting on the breast and colon. Estrogen, the female hormone, is a promoter of breast and uterus cells.

Even some mechanical processes such as abrasions or other wounds can act as promoters. For example, when the skin of a mouse is painted with a carcinogen and then cut with a knife, a row of tiny tumors often sprouts along the healed wound. The process of wound healing is, of course, one in which normal cell proliferation is speeded up dramatically, usually as the result of the release of various growth stimulants by special cells at the site. Normal cells respond and then stop when the wound is healed. Cells that have been genetically transformed do not know when to stop.

Once again, however, there is a mechanism that appears able to block the cancer process, at least insofar as it depends on promoters. Several chemicals found in a normal balanced diet appear to be able to block the effects of promoter molecules. Some of these chemicals, such as the vitamin A derivatives called retinoids, are metabolic byproducts of many dark green or yellow vegetables. Animals fed diets rich in retinoids appear to be unusually resistant to the tumor-causing effects of promoters, even if they have been exposed to a carcinogen. Many researchers speculate that people who eat diets rich in these foods may be tipping in their favor the balance of pro- and anticancer molecules in their cells.

But if the balance tips toward the cancer process, the cells continue accumulating mutations until they get a most fateful one, a mutation that causes the cell to keep dividing after the promoter substance is removed. It is a mutation that, in effect, jams the accelerator to the floor and ignores all stop signs. The result is an exponentially growing mass of cells, all descendants of the one that gained the key mutation that makes a promoter unnecessary and all genetically geared to keep proliferating. The tumor grows ever larger, at first microscopic in size but eventually one that cannot be ignored. By the time it is detected, a typical tumor may have more than a hundred million cells.

Not all cancerous mutations are the result of chemical carcinogens. Some are caused by radiation, from the ultraviolet of sunlight that penetrates only a millimeter into the skin to the gamma rays of radioactive decay that can penetrate the whole body and hit a gene anywhere. Still other mutations may be the result of random rearrangements of genes or gene segments along the chromosome. Sometimes during cell division pieces of a chromosome may break off and get spliced back into the wrong place. Jorge Yunis of the University of Minnesota has shown that just such a translocation, as the phenomenon is called, can be detected in tumor cells from nearly all victims of Burkitt's lymphoma. Moreover, the break point seems to be the same in most cases—a piece of the end of chromosome 8 swaps positions with a piece from the end of chromosome 14. As it happens, chromosome 8 carries a potential oncogene, or proto-oncogene, very near the break point. The translocation brings the proto-oncogene into the sphere of influence of a regulatory sequence that drives the cell to make antibodies. Since these are cells of the immune system whose job is, indeed, to make antibodies, the regulatory sequence is highly active. The only problem is that its effect now extends to a gene that is supposed to be inactive.

Whatever the mechanism that leads to cancer, the result of enough mutations of the right kind is a tumor, though at this point it may be only of the kind called—sometimes only euphemistically—benign. It is simply a mass of cells that grows steadily larger but does not invade surrounding tissues or spread to distant parts of the body. The tumor's cells must acquire several more mutations to endow them with the powers to do those things, the powers that make them malignant. We'll return to this matter later in this chapter.

Up to this point, we have described cancer as a genetic disease, a result of derangements in the DNA. This is the current view, but that understanding, unlike much of the other information in this book, was won only very recently. The first solid clues emerged in 1975, and the beginning of a coherent theory of cancer did not reach its full form until the 1990s.

Of course, the trail started much earlier. In 1911 researchers discovered the existence of a virus that caused cancer in the muscle cells of chickens. There the matter sat, quite apart from the mainstream of research on cancer, until 1970 when scientists found that the chicken virus carries one particular gene that if placed inside a normal, slowly multiplying cell, turns it into a wildly proliferating cancer cell. The gene was dubbed an oncogene, from the Greek *onkos,* meaning "a mass." In 1975, the American researchers Harold Varmus and J. Michael Bishop discovered that a gene almost identical to the chicken virus gene existed in normal cells of animals and humans. That finding, which won a 1989 Nobel prize for the pair, began to shape the modern view of cancer. The exact gene that exists in humans (and all animals) doesn't cause cancer, but if it undergoes a certain, tiny mutation, it acquires the DNA sequence of the cancer-causing oncogene and turns the cell cancerous. Since that discovery, researchers have found more than seventy other genes that, if mutated, become oncogenes.

Although the exact role played by every gene is not known, it is clear that the genes must be playing some role in the cancer process. That conclusion has been affirmed over and over again by experiments in which copies of the oncogenes are injected into cultured normal cells. With each added oncogene, the cell takes on some new ability typical of cancer cells. The injected cell differs in the way it looks and acts, and it passes on these differences to its offspring. If the altered cell is injected into a laboratory animal, the animal gets cancer.

Another clue to the fundamental importance of proto-oncogenes is that each type can be found in all forms of life from yeasts to human beings and in some cases even in plants. The fact that these genes have changed little in billions of years of evolution suggests they play critically important roles in the maintenance of life itself, regardless of the cell's particular job in any given species. Think, for example, of the genes whose protein products help regulate the cell cycle. Such a protein might tell the cell to divide under certain circumstances. Imagine, now, that the gene is damaged so that its protein no longer waits for some outside signal but always tells the cell to divide. Just such an oncogene, called *ras,* has been found in a considerable number of human tumor cells. The normal form of the protein resides just inside the cell membrane and has the characteristics of the molecules whose job is to relay signals brought by proteins arriving at receptors on the cell surface. It looks as if the mutant ras simply relays a signal even when nothing has arrived at the receptor.

Several proto-oncogenes, to cite other roles, contain codes for enzymes that attach phosphates to specific sites on proteins. This process, called phosphorylation, is one of the most common and most powerful regulatory mechanisms in cells. When certain proteins are phosphorylated, they change their shape and their biochemical powers. When the same proteins are dephosphorylated, the shapes and powers change back. As we have seen in earlier chapters, many of the metabolic steps essential to life, including several of the fundamental housekeeping mechanisms common to all cells, are governed through this process. Many oncogenes, it turns out, are genes for enzymes that phosphorylate various specific proteins. Such molecules are called protein kinases. Since a single type of protein kinase may phosphorylate several other types of molecules within the cell, a single mutation could have wide-ranging effects throughout the cell.

"One enzyme, by phosphorylating a number of proteins, can vastly alter the functioning of a cell," says Bishop. "Cancer genes may not be unwanted guests but essential constituents of the cell's genetic apparatus, betraying the cell only when their structure or control is disturbed by carcinogens."

One unusual disturbance by carcinogens can be suffered by the gene that normally codes for the EGF receptor. This is a molecule that normally resides in the cell's outer membrane, holding out a receptor for a substance called epidermal growth factor (EGF), which, despite its name, is required by many types of cells to undergo cell division. When a molecule of EGF comes along and binds to the receptor, the receptor's intracellular tail changes shape in a way that makes it a protein kinase. It starts phosphorylating certain other proteins, thus helping to drive the cell to divide. As it happens, the EGF receptor gene is a proto-oncogene. It can be mutated in a way that leaves the resulting protein shorn of the part that acts as the receptor. As a result, the receptor's intracellular tail automatically assumes the shape that makes it a protein kinase, and it perpetually sends the chemical signals to divide. The damaged EGF gene is now an oncogene.

More recently, molecular biologists have found a different kind of cancer gene, one whose history can be much like that of an oncogene but whose normal role is to keep cell division under proper control. If oncogenes are the accelerator pedals of cancer, these are genes whose protein products keep the brakes on cell proliferation. If one of these genes is damaged, the brakes are released and the cell automatically leaps into high gear. Such genes are called tumor suppressor genes. If anything, they are even more important players than proto-oncogenes in human cancers. The best-known such gene, dubbed *p53*, is known to play a crucial role in about half of all human cancers. It is commonly implicated, for example, in breast cancer, lung cancer, colon cancer, and several other common malignancies.

Although about a dozen other tumor suppressor genes are known—including the recently discovered breast cancer susceptibility gene (called *BRCA1*)—*p53* is, by far, the best understood. Incidentally, the p, according to the standard nomenclature for these things, stands for protein and the 53 for its molecular weight in units of 1,000 daltons. (A dalton is the same unit that an earlier generation, prior to the decision to honor chemistry pioneer John Dalton, referred to as molecular weight.) Sometimes the designation is confusing because p53 is used alone to refer both to the protein and to the gene that encodes the protein. The p53 mechanism is also the most completely understood story linking the newest knowledge of the cell-division cycle, DNA mutations, and the phenomenon of programmed cell death, or apoptosis. In each of these fields, independent teams of researchers had turned up different clues that all led to the story of p53. Here, in a nutshell, is what is now known.

One of p53's several jobs is to help regulate a cell's progress through the cell-division cycle. Specifically, the protein's role is to stimulate the DNA inspection and repair enzymes and to prevent a cell from replicating its chromosomes (the first major step in preparing to divide) until all necessary repairs are complete. If that happens, p53 signs off on that process and allows the next step—the copying of the chromosomes—to proceed. If the damage is too severe, however, p53 somehow senses the failure of DNA repair and commands the cell immediately to kill itself—to undergo programmed cell death. The p53 protein, then, is a quality control agent with responsibility not to its own cell but to the republic of cells. Its core responsibility is to keep a cell with damaged DNA not only from proliferating but from continuing to exist at all.

How does the p53 protein do this? The whole process is not yet understood, but p53 is known to be a DNA-binding protein that attaches itself to several specific genes to govern their activities.

Imagine, then, what would happen to a cell if its *p53* gene were to be mutated. The quality control agent would be eliminated. If the cell's genes became mutated, the cell would not be weeded out; it would be permitted to divide, increasing the population of genetically defective cells. Without p53's ability to stimulate DNA repair, without its role as a checkpoint in the cell-division cycle, without its ability to trigger apoptosis, the future of such a genetically damaged cell cannot be good. The cell is now free to accumulate further mutations—changing proto-oncogenes, for example, into full-fledged oncogenes—that will endow it with cancerous abilities.

So sweeping are p53's powers that scientists have been able to "cure" cancer cells growing uncontrollably in a dish by injecting good copies of the *p53* gene into them. Even malignant cancer cells that have several mutated oncogenes instantly revert to normalcy when given a good *p53* gene. If a way could be found to slip normal *p53* genes into each of the billions of cancer cells in a human patient, the disease might be halted. But for obvious reasons that will be very difficult, if not impossible.

Exactly what does a specific carcinogen do to a specific cancer gene to turn it bad?

In 1991 cancer researchers got their first clear look at one example. Until then, although the term "carcinogen" had become a household word, the case for the existence of such substances rested almost entirely on circumstantial evidence. For example, since heavy smokers got most of the lung cancer, it seemed obvious that tobacco smoke contained cancer-causing chemicals. And when researchers applied suspected carcinogens and promoters from the smoke onto the backs of laboratory mice, tumors sprouted only under the applied substance.

The circumstantial case was quite solid. Indeed, no reasonable person doubted that certain chemicals somehow wrought a profound change, turning normal cells cancerous. But still, nobody knew the precise mecha-nism by which a carcinogen actually deranged a cell.

That question has been answered, at least for the earliest step in the path toward liver cancer, which is one of the world's top five cancer killers and is especially prevalent in Africa and Asia, where victims tend to be young and middle-aged, but relatively less common in the United States, where victims are usually older. Still, wherever liver cancer strikes, it is one of the most vicious cancer killers of all, rarely responding to treatment and usually bringing death within weeks or a few months of the first symptoms.

In 1991 two scientific teams—working separately but with knowledge of each other's work—found exactly the same mutation occurring in the same codon within the same gene. More importantly, the two groups showed that the mutation was caused by the same carcinogen—a toxin produced by a fungus that grows in corn, peanuts, and certain other foods. It is called aflatoxin. This was the first time scientists proved conclusively that a suspected carcinogen not only damages a cell's DNA but that the damage is the same almost every time.

So specific is the damage that one of the scientists involved suggests the discovery may open the door to a new science that might be called mo-lecular forensics. For example, it may someday be possible to examine cancer cells, look for the damage, and say which carcinogen caused it. Today, with few exceptions, nobody can say what caused any given case of cancer. If other carcinogens turn out, like aflatoxin, to inflict a specific kind of recognizable genetic damage, a patient could know that his cancer was caused by a certain pesticide, or by the hydrocarbons in charred meat, or what have you. It would be little comfort for the patient, but it would be invaluable for public health officials trying to figure out which chemicals to keep out of the environment.

The gene mutated by aflatoxin is none other than *p53*, the tumor suppressor gene. The previously known instances of *p53* mutation in many different kinds of cancer involved alterations in scores of different codons, and it was impossible to relate the mutation to any particular carcinogen, or any other cause. There was no way to tell what might have caused them. That's what makes the new work on *p53* and liver cancer so important.

In half the liver cancer patients studied by the two groups, the *p53* genes were all mutated at the third base in Codon 249. Normally there should be a G there. All the mutant forms but one had a T instead. This meant that when the cell followed the gene's instructions, it installed the wrong amino acid in the 249th position. This meant the 249th amino acid in the protein, which should have been arginine, was actually serine. The link seems all the more solid inasmuch as laboratory tests on cells show that aflatoxin causes G-to-T mutations in many genes.

Never before had scientists seen such a consistent mutation. Adding to the wonder was the fact that one group of patients was in South Africa and the other was in China. Both regions are hotbeds for liver cancer, and it is one of the leading killers in both places. The two regions had also been linked by epidemiologists looking for a common explanation for the high liver cancer rates. Aflatoxin is common in both regions and exposure to aflatoxin had been a well-established risk factor for liver cancer.

As it happens, a second risk factor for liver cancer—not just in Africa and China but in the United States as well—is hepatitis B, a viral infection that damages liver tissue. According to epidemiological evidence from Third World countries, both aflatoxin and hepatitis are implicated. But with the new genetic findings, a more detailed scenario can now be envisioned.

Since hepatitis virus kills liver cells, the surviving cells try to repair the damage by proliferating at a faster than normal rate. This means the odds are increased that a mutation will arise, since errors are most likely during the step in which the chromosomes are being duplicated. A molecule that happened to interfere with DNA replication could cause a mutation at this point. If the speculations are correct, aflatoxin has some chemical proclivity that causes it to interact with Codon 249 while the *p53* gene is being copied. One possible scenario, then, is that hepatitis spurs liver cells to divide more rapidly (making it a promoter), thus increasing the odds that an aflatoxin molecule will slip a monkey wrench into the DNA-copying works. From that subtle change, the researchers have established, proceeds one of the deadliest of all cancers.

One of the most intriguing new findings about how cancer cells reacquire the powers of their free-living primordial ancestors involves a special type of DNA sequence—TTAGGG repeated many times—found at the tips of all chromosomes. This DNA coding, called a telomere, is not the recipe of a gene, nor a regulatory sequence. It is there to compensate for what might be called a design flaw in DNA polymerase, the molecule that copies chromosomes for cell division. When the huge polymerase molecule works its way along an unzipped DNA strand, it needs to grasp the chain on both sides of the region it is copying, rather like a person eating corn on the cob. But when DNA polymerase comes to the tip of a strand, one "hand" must let go before the last of the bases can be duplicated. When the DNA polymerase molecule drops away from the strand it was working on, the last fifty to sixty-five bases of the old chromosome remain unduplicated. The new DNA strand is not, after all, a perfect copy of the old one.

In most normal cells of the body, bits of telomeric DNA are lost each time a cell divides. This usually causes no problem because all chromosomes have several thousand extra bases capping their tips. But when the last of the protective sequences is gone, subsequent cycles of cell division eat into genes, producing daughter chromosomes in which the last gene will have lost some of its codons. The daughter cells that inherit these chromosomes will be deprived of these genes. The new cells become badly deranged and may die, depriving the body of their function. This process is thought to play a role in aging, a topic we'll visit in the next chapter.

Cancer cells have escaped this death sentence. Calvin B. Harley and colleagues at Geron Corporation, a biotechnology firm in Menlo Park, California, have found that cancer cells make an enzyme that rebuilds telomeres at each round of cell division. DNA polymerase loses a few bases, but before mitosis proceeds, an enzyme called telomerase grabs the new DNA strand and, because it figuratively needs only one hand to work, synthesizes a few repeats of TTAGGG to make good the loss. Telomerase can do this because it is a complex of a protein and a short segment of RNA that carries the template for making new DNA. Harley believes that this is a key source of the cancer cell's immortality.

Where do the renegade cells get their telomerase? The gene for it is right there in their nuclei. All cells have a telomerase gene, but it is supposed to stay dormant except in cells of the gonads. The cells that make eggs and sperm need protection from telomere loss because they give rise to the founding cells of the next generation, which must start off with full-length telomeres. Harley and his group have tested more than eighty samples of human cancer cells and found that all had switched on their telomerase genes.

Harley has found that cancer cells growing in culture, the famous HeLa cells mentioned in Chapter 1, can be made to die if they are simply deprived of telomerase. The researchers were able to kill the cells by giving them an artificial gene that produces a messenger RNA with a sequence exactly complementary to that of the messenger RNA needed to make telomerase. The artificial RNA, called antisense, binds to the "sense" form of telomerase RNA, locking the two strands so that neither can proceed to guide protein synthesis.

While this approach is highly impractical for use on patients, it shows that if cancer cells can be deprived of their telomerase, they will die. Geron and many other laboratories are looking for drugs that will have a comparable effect, getting into tumor cells and fouling up either the telomerase gene or its product.

Even if a cell breaks the shackles that keep its rate of division under control, however, it is not yet a malignant cancer. Mere uncontrolled cell division, for one thing, produces "only" a benign tumor. Benign tumors grow as a single lump, every cell in the lump a direct descendant of the founding cell that first acquired the right mutations. Contrary to their name, benign tumors can be dangerous if, for example, the lump grows large enough to block off some vital passageway such as the trachea or a blood vessel or the digestive tract. But if found before they cause trouble, benign tumors can be surgically removed and the patient totally cured. Cure is assured because the cells of benign tumors lack the ability to invade and damage other tissues or to break away from their original site and wander off to new parts of the body. It is these powers that make a cancer malignant.

Malignant cells have a varied behavioral repertoire. Instead of simply enlarging within its native tissue, a malignant tumor may sprout tentacles that push into surrounding tissues. Wedges of tumor cells, displaying an ability quite alien to their former role as subservient cells of, say, the lung or the breast, secrete special enzymes that destroy surrounding normal cells. Hippocrates was the first on record to recognize this difference between benign and malignant tumors. The invasive wedges so reminded him of crab claws that he gave the disease the Greek name for the crab, *karkinos*. In English, the term survives as *carcinoma* and as the Latin word for "crab," *cancer*.

The fact that malignant cells invade and benign cells don't is the result of the malignant cell having additional genes at work. For a tumor to cut through its surrounding tissue, it must use special enzymes that can dissolve the extracellular material that fills the spaces between normal cells. Additionally, the cancer cell must also "chew" its way through the so-called basement membrane that encloses tissues. The enzymes that can do the job are not foreign to the cell. They are the same type of enzymes that are essential to the early development of mammalian embryos. Having emerged from the fallopian tube into the uterus, the blastocyst secretes such enzymes to invade the wall of the uterus and implant itself. At other times during embryonic development, certain cells migrating from one part of the embryo to another will produce minute amounts of these enzymes to open paths for themselves. Once development is complete, however, the genes for these enzymes are supposed to turn dormant and stay that way. Somehow cancer cells find these long disused genes and turn them on again.

If tumors are to grow larger than about two millimeters, they must develop yet another special behavior. They must send out chemical signals that beckon to nearby blood vessels, calling them to sprout capillaries that will snake into the tumor to deliver oxygen and nutrients and take away metabolic waste products. For all their alien nature, tumor cells still have the same basic requirements for life as any other cell. Once again, they achieve this goal by invoking a cellular ability that is normally supposed to operate only at some other time and place. Blood vessel growth is most common during development, causing the circulatory system to ramify throughout the growing body. During this period, cells secrete so-called angiogenesis factors ("angio" from the Greek for "vessel") that diffuse outward until they reach a cell forming a wall of the nearest blood vessel. The cell responds by sprouting an arm that reaches toward the source of the signal and then divides, producing another cell that stretches still further toward the oxygen-starved source of the angiogenesis factor. As the newly forming capillary lengthens, its cells form huge hollow spaces that link up, opening a channel from cell to cell through which blood can flow. New blood vessels also form during wound healing and with the growth of fat layers in the body, and in women of reproductive age, the process cycles monthly to build up the lining of the uterus before it is shed in menstruation. Tumor cells reactivate this process to keep their rapidly growing mass equally well served by the circulatory system.

Because cancer is almost the only place angiogenesis occurs in an otherwise uninjured adult, cancer researchers, led by Harvard's Judah Folkman, have long seen it a potential target for trying to block the growth of cancer. Indeed, several substances that block angiogenesis have been discovered. In animals with cancer, these chemicals dramatically inhibit the growth of tumors and prolong survival. Work is under way to test their effectiveness in human patients.

Even the ability to recruit blood vessels and to invade surrounding tissues would not make the malignant cancer so dangerous if it were not for still other unusual powers of the renegade cells. When cancer kills, it is most often because of the malignant tumor's ability to dispatch individual cells, sending them out to establish new colonies in other parts of the body. This process, known as metastasis, requires that the tumor cell somehow break the junctions that held it to surrounding cells and then synthesize enzymes and other substances that, in effect, dissolve a hole in the wall of a blood vessel. The escaped malignant cell then slips into the blood vessel and drifts with the current.

When the aimless voyager happens to lodge in some narrow capillary or in the bend of a sharp turn, it produces enzymes that chew its way out of the blood vessel and into whatever tissue it happens to be in. If the cell is suited for its new environment, it resumes dividing again and again, creating a new tumor that may invade local tissues and, eventually, send out its own roaming cells. Some human tumors are estimated to dispatch millions of metastatic cells daily.

The process of cancer cells acquiring more and more powers is known to physicians as progression. In recent years, cancer biologists have realized that progression is the cellular equivalent of evolution by natural selection. All living things are the result of a process in which genes are mutated at random in the parent's generation and the offspring are thrust into the natural environment where the differences caused by those mutations may prove deleterious or, in rare instances, advantageous. If the new trait (governed by the changed gene) improves the individual's odds of surviving in a given environment, it is likely to leave more surviving offspring than its unmutated siblings. The environment is said to have selected the new form of individual for survival. And because the mutated gene is passed on to yet another generation, those individuals, too, will have enhanced prospects for survival. Eventually, the mutant version may become the predominant example of the species. But the process of mutation and natural selection continues. Over time, many changes can accumulate, creating a new species that is dramatically different from its distant ancestors.

Exactly the same thing happens to cancer cells, the physical conditions of the body serving as the cells' natural environment. First may come the ability to keep dividing when other cells stop. Then may come the ability to invade neighboring tissues, then to break away from the original tumor, then to survive in a different environment somewhere else in the body. Early in the progression/evolution of many tumors, it now appears, comes the inactivation of the *p53* gene, the one that lets cells continue dividing even if some of their DNA is damaged. That mutation—and others affecting less well-known genes that serve as checkpoints in the cell-division cycle—makes the cell genetically unstable, makes it prone to accumulate mutations.

As a result, cancer cells evolve faster than organisms in nature under normal genetic controls. They throw out many new gene sequences and even radical chromosome rearrangements to be tested in the varied environments within the body. In a sense, cancer cells evolve into something that is very different from a normal human cell, changing genes and regulatory sequences and even hosting such gross alterations as the number of chromosomes per cell. Cancer cells evolve such fundamental genetic differences that, by the standards of ordinary biology, they would be classified as very distantly related species. In fact, the derangements in cell-cycle controls in cancer cells are so profound that they depart from the standard processes that operate in identical fashion not only in all animal species but among all cells with nuclei, from yeasts to humans. This makes cancer cells a most alien life-form indeed.

Unfortunately, one of the most powerful forms of natural selection on cancer cells is often exercised by giving patients anticancer drugs, the so-called chemotherapy. The drug may kill a large number of cancer cells, but if even one is somehow resistant, it will now be able to show its ability to thrive in the new, drug-contaminated environment. Unimpeded by competition for food and oxygen with its other cancer cell kin, the drug-resistant cell can now multiply freely. If it is present when chemotherapy is tried, a relapse, or "recurrence," is all but inevitable. As it happens, many cancer drugs, cell biologists now know, kill by inducing apoptosis. So one way for a cancer cell to be resistant is to have undergone mutations that wrecked the genes needed to carry out programmed cell death.

Early in the cancer cells' evolution, the body has a chance to mount a defense. The immune system may recognize the cancer cells as alien and destroy them before they can grow or find new homes in the body. Because tumor cells have abnormal proteins on their surfaces (the products of mutant genes), the cells of the immune system may generate antibodies that bind to them. The immune system's natural killer cells seem to make cancer cells a specialty. When they find one, they sidle up next to it and release enzymes that open holes in the tumor cell's membrane. The fearsome cell simply leaks to death. Since natural killer cells wander the body outside the bloodstream, they may even destroy original cancer cells before they can form tumors. Some cancer researchers think it is fairly common that cancer cells arise in the body but that the natural killers dispatch them before they can cause any trouble. Only when the immune system is somehow crippled, they suggest, do tumors develop to the danger point.

Often enough, unfortunately, the malignant cells do evade the cellular police. Soon after having left their site of origin, the cells take up residence in distant tissues and resume multiplying, creating new tumors that eventually will begin shedding metastatic cells of their own, all armed with a coordinated set of cancer genes. Eventually the one original cancer cell and its legion of descendants will travel throughout the body, establishing colonies in distant tissues, monopolizing the food supplies, proliferating as if immortal.

If removed from the body and put in a cell culture flask with a food supply, tumor cells do indeed seem immortal. They are the founders of all of the longest lived cell lines in laboratories, many, like the HeLa cells of

Henrietta Lacks who died in 1951, are still going strong, long after they killed the human body that first gave them life.

Cancer cells have broken the compact established billions of years ago when the first free-living, one-celled organisms banded together to form multicelled organisms, to form the republic of cells. The spectacular evolutionary success of those multicelled organisms has always, however, remained under the threat that some cells would rebel. Indeed, cancer has probably always been a fact of multicellular life. Fossilized bone tumors have been found in dinosaur skeletons. Today we know that the original compact of multicellularity depended on some fragile genetic links and that cells remain all too able to break those ties to proliferate selfishly.

Perhaps the Book of Genesis says that when we /adam ate of the Tree of Knowledge of Good & Evil, death was introduced → perhaps it was moving from the immortal cellular stage of evolution to one where the cells do die.

**D**eath, contrary to conventional wisdom, is not an essential part of life. Some forms of life are immortal. They never die.

Take, for example, organisms that consist of a single cell. Unless they encounter the microbial equivalent of getting hit by a truck, many of them never encounter death. After bacterial cells or many kinds of protozoans have lived for a while, they divide, and presto, they are instantly reincarnated as two youthful individuals. The spark of life is passed to a new generation, and yet the old generation leaves no corpse. The contractile ring that cinches tighter to pinch the aged parent cell in two transforms one unit of life into two. Each daughter cell incorporates roughly half the living matter that made up the parent. Then the new little cells feed and grow larger. Eventually the cell-cycle clock tolls again and the daughters repeat the death-defying feat of cell division. If a microbial cell ages in any concrete way (it is not clear that it does), it is rejuvenated in the course of cell division.

Through repeated cycles of growth and division, life continually battles death, extending its power to convert more of the surrounding world into ever more copies of itself, copies of the molecular structures of DNA and the enzymes, the proteins and the membranes that make for life. The small daughter cells take in some of the world around them (mostly in the form of nutrients) and reorganize it to become part of their living selves. We are, even on the cellular level, what we eat.

Like a crystal whose geometry causes newly arriving atoms to link up in predictable patterns, the molecular structures of life dictate the architecture of newly incorporated matter. Thus, when the cell divides, the thing that is transmitted to each daughter—the heart of the ability to escape death—is a seed of the architecture of life, with inherent powers to organize still more nonliving matter as each new cell grows and divides.

The power of life to enlarge itself is so extraordinary that it seems to defy one of the fundamental laws of physics, the Second Law of Thermodynamics. The Second Law is the one that says everything within a closed system eventually falls apart—unless you keep importing new energy from outside, in which case it is not a closed system. The law says that every time work gets done, some of the energy is lost, usually in the form of heat, and that the energy can never be recaptured. The law dictates that the fate of the universe is to run down, disintegrate, and become a homogeneous, cosmic soup. The Second Law establishes the famous concept of entropy, which is a physicist's word for disorder. As time goes by and things happen, all the order of the universe—including the discrete objects such as stars and planets—will break down. Stars will burn out and either blow apart or cool to cinders. Planets will be ripped to smithereens by the gravitational forces of close encounters and the impacts of outright collisions. The heat of the stars will be poured into open space, most of it never to be recaptured. Entropy rules, the physicists tell us. Entropy is ultimate death.

But two antientropic phenomena that build order out of chaos are evident in the everyday world—crystallization and life. Both can diminish entropy only because they consume energy from outside their own systems. Order cannot emerge from disorder unless some organizing force is in operation, and such forces require energy.

A lump of sugar easily dissolves in a cup of tea, surrendering the order of its crystalline lattice to entropy. The crystals cannot reassemble themselves from the solution. But if outside energy is supplied, in the form of heat, for example, the liquid can be evaporated away, leaving the sugar molecules, which do not turn to vapor, to recrystallize in the bottom of the cup. Under these conditions, each molecule of sugar "tells" its neighbor what position to assume in the growing crystal. The instructions are inherent in the structure of each sugar molecule. The entropic disorder of the dissolved sugar is reversed. The result is a crystal of predictable symmetry—cubic for sugar, for example, or hexagonal for water (if energy is used to remove heat from the water).

Life is like crystallization, only with far more spectacular and more profound results. The molecular structures of a living cell can, like crystals, extend their pattern of organization to more and more matter as the cell grows. As the cycles of growth and division replay themselves again and again, each form of life seeks to convert more and more of the world around it into its biomass and that of its progeny. (Of course, when life first arose on Earth the total biomass must have grown rapidly as the earliest cells multiplied without competition, feeding on nonliving foods in a virgin world. Nowadays, Earth's total biomass probably changes little since most ecological niches have long since been filled and so many species now try to grow by feeding on one another.) Like crystallization, life is antientropic because it is not a closed system; it can import energy from outside the system. The most abundant form of energy is, of course, sunlight, captured by plants. Almost all life on Earth then simply reprocesses the energy stored in plants. (The chief exceptions are the animals that live around the deep-sea hydrothermal vents, where the bottom of the food chain is the chemical energy in the inorganic compounds that come out of the vent.)

The concept that the phenomenon of life emerges from the architectural arrangements within and between molecules has emerged quite slowly in biology. A key part of that emergence began with a controversy that raged for centuries about the nature of death and resurrection—a problem as much cultural as scientific—that was settled only in recent years. The story of that controversy is one of the least appreciated in biology but one of the most instructive.

It involved some rather small animals whose survival depended on their astonishing ability to enter into a state of being that conventional biology long defined as death. When they need to, these common but little-known life-forms simply dry up, shriveling into wrinkled kernels. In this state they do not eat or breathe. They do not move. Digestion stops. Nervous systems shut down. All metabolism ceases. For periods of weeks, months, and even years, the little kernels simply endure.

But with the return of moisture, they revive. Within minutes, the dessicated kernels absorb enough water to swell back to their normal proportions, and all of the hundreds of biochemical processes that make for life as it is conventionally defined begin again. Shrunken lumps of animals that may have spent ten years in a bottle on a laboratory shelf virtually spring back into action when wetted, wriggling and writhing in a Petri dish, or literally crawling off the microscope slide in search of food. Like the frozen cells in the tank farm at the American Type Culture Collection, these tiny organisms exist in a limbo that challenges the usual definition of life.

Most biology books teach that life cannot exist without water. Most biology books say that life is characterized by the various chemical processes that are summed up in the word "metabolism." And most biology books say that when metabolism ceases, death ensues.

Most biology books, unfortunately, are wrong. Or, at least, they neglect an astonishing biological ability, a natural form of suspended animation, possessed by scores of species that can be found by the billions in almost every habitat on Earth. These include such soil-dwelling animals as the wormlike nematodes and the insectlike tardigrades, and aquatic species like the vase-shaped rotifers of freshwater ponds and the brine shrimp that live in salt lakes (Figure 13.1). Although small—most are less than a millimeter in length—they are fairly complex, multicellular animals equipped with digestive tracts, reproductive organs, nervous systems, muscles, and other specialized structures and tissues that make them more closely related to higher forms of life than to bacteria or protozoa.

These animals live in a wide variety of hostile environments, such as the thin soils of the Arctic tundra, where they remain frozen most of the

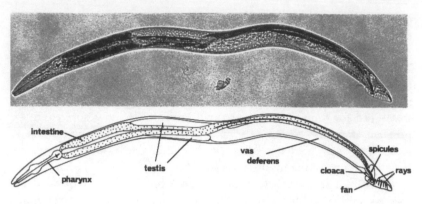

**Figure 13.1**  A nematode, one of several species of small but complex animals that can undergo suspended animation by drying out under proper circumstances. Dried forms of these "cryptobiotic" animals may persist for years without metabolizing and then, when wetted again, resume normal activity. (From J. E. Sulston and H. R. Horvitz, *Developmental Biology* 56 [1977]: 110–156. Photo courtesy of Professor John Sulston)

time, and in hot desert sands, where they receive so little moisture that they are active only a few days each year. They endure in sun-baked mud, ready to resume activity when rain creates brief-lived puddles. And they inhabit virtually all other soils as well, from loamy forest floors to suburban back yards to the thin organic layer that forms as mosses creep up the north sides of trees and buildings.

These ubiquitous animals pose a fundamental challenge to the conventional definition of life. According to that definition, these organisms, when they exhibit no sign of metabolism, are not alive. But are they then dead? When moistened again, is life spontaneously generated? To avoid such conceptual and semantic dilemmas, the few biologists who have studied the phenomenon generally refer to it as cryptobiosis, meaning "hidden life."

So strong was the traditional metabolism-centered view of life that until recently most biologists suspected that cryptobiotic animals were not totally inactive. There was good reason to think that at least a little water remained inside the driest specimens, bound to the organelles inside the cells. If that were the case, metabolism might be continuing at a rate too slow to be detected. After all, many higher animals can reduce their metabolic rates by hibernating in winter, and many others can cut metabolism even lower by estivating in dry summer heat. Cryptobiotic animals, many researchers suspected, were simply extending a familiar capacity to a previously unknown extreme.

Recently, however, it has been established that although a few water molecules do remain even in the driest organisms, they constitute only a small fraction of the minimum needed for any of the biochemical interactions of metabolism to happen. Most proteins, for example, must be awash in water to assume their functional three-dimensional shape. Tardigrades and nematodes, like most animals, are normally 80 percent to 90 percent water. In the cryptobiotic state, the organisms are only about 3 percent to 5 percent water. In laboratory conditions some have had their water content reduced to 0.05 percent and were then revived. Most authorities now agree that no metabolism takes place during cryptobiosis. The term no longer means "a hidden form of ordinary life" but rather "another form of life; a state of being in which the active processes of life are latent or unexpressed."

In the cryptobiotic state, all that remains of a living organism is its structural integrity. A dry nematode, tardigrade, or rotifer may be shrunken and deformed, but it maintains all the basic connections that keep the molecules and larger structures of its cells together and that keep the cells in their proper places within the organism. Just add water, and the proteins and nucleic acids simply regain the ability to flex and bump and interact according to the chemical proclivities implicit in their structure. Life, to recapitulate Chapter 1, is not the result of some mystical animating force that inhabits cells. Life is nothing more—and nothing less—than the structural arrangement of certain molecules that behave chemically in specific ways when in the lubricating presence of water. Water also does more than lubricate; it participates in many chemical reactions, and its electromagnetic properties influence both the courses of chemical reactions and the shapes of intracellular structures.

Such an appreciation of life was very far from the mind of Antoni van Leeuwenhoek, the pioneer microscopist, when he wrote in 1702 from his native Amsterdam to the Royal Society of London about "certain animalcules found in the sediment on the roofs of houses." Leeuwenhoek found that the animals, probably rotifers, that dried out while under his microscope appeared to be dead but that if he put a drop of water on them, they soon revived and began swimming and crawling. Leeuwenhoek's report was soon forgotten, and 40 years later the phenomenon was reported anew, starting a controversy that would not subside for 120 years. In 1743 John Turberville Needham, a Jesuit scholar, rediscovered cryptobiosis, this time in the nematodes called eel worms that grow inside the swollen, cystlike galls of blighted wheat plants.

"We find an instance here," Needham wrote, "that Life may be suspended and seemingly destroyed; that by an Exhalation of the Fluids necessary to a living Animal, the Circulations may cease, all the Organs and Vessels of the Body may be shrunk up, dried and hardened; and yet, after a long while, Life may begin anew to actuate the same Body and all the animal Motions and Faculties may be restored, merely by replenishing the Organs and Vessels with a fresh supply of Fluid."

Needham's discovery ignited a wide-ranging debate as to whether resurrection, as the phenomenon was often called, was something that occurred only once, as the church held, or whether it was a common phenomenon exhibited by even the lowliest of animals. The debate spread to Italy and attracted the attention of Lazzaro Spallanzani, a renowned professor at the University of Pavia. Uncharacteristically, Spallanzani, without trying to replicate Needham's experiments, dismissed the discovery. Fortunately, experiments by others showed that dry nematodes do revive. Eventually, Spallanzani could no longer ignore the subject and began his own studies. He quickly repented his earlier hastiness, confirmed Needham's findings, and extended them. He found that dried nematodes could withstand extremes of temperature that would kill the same animal in the active state. Moreover, Spallanzani's dried animals revived after long periods in a vacuum, and they survived otherwise fatal electric shocks.

"An animal which revives after death," Spallanzani wrote, "and which, within certain limits revives as often as we please, is a phenomenon as incredible as it seems improbable and paradoxical. It confounds the most accepted ideas of animality; it creates new ideas and becomes an object no less interesting to the researches of the naturalist than to the speculation of the profound metaphysician."

Indeed, with Spallanzani now firmly in the "resurrectionist" camp, the churchly metaphysicians had a more powerful adversary than ever, and they fought back. Their chief argument was that dried animals had not really ceased their life processes but had only suppressed them below the limits of detection. It was a reasonable objection, given the inability at that time to demonstrate whether such tiny creatures were or were not metabolizing.

The debate, pitting the resurrectionists against the antiresurrectionists, continued until the middle of the nineteenth century. In France, reviving dry nematodes under microscopes became a popular entertainment in fashionable salons. By the late 1850s the controversy had come to a head, and the leaders of each side asked that country's prestigious Biological Society to determine officially whether nematodes, rotifers, and tardigrades ceased metabolizing when dried out. The French society appointed a commission of seven distinguished scientists headed by Paul Broca, the great anthropologist and anatomist. The Broca commission conducted experiments, held forty-two meetings, and produced a six-thousand-word report that remained the last word on the subject until just three decades ago. The distinguished scientists dried rotifers in a vacuum for eighty-two days, heated them to the boiling point of water (though the creatures were kept dry) for half an hour, then revived them nicely with but one drop of room-temperature water. Nonetheless, the commission felt it was unable to prove conclusively that all life processes had totally stopped. So the commission came down squarely on the fence, neither affirming nor denying either side.

Cryptobiosis research has continued off and on ever since. Rotifers and tardigrades have now been heated to more than 300 degrees Fahrenheit for several minutes; they have been cooled almost to absolute zero, that is, to nearly 459 degrees below zero Fahrenheit. In both cases they revived. Dry nematodes and tardigrades have survived two full days in a vacuum nearly as airless and low in pressure as that of outer space. Tardigrades have withstood an astonishing 570,000 roentgens of ionizing radiation for twenty-four hours—more than a thousand times the lethal dose for a human.

Just as such experiments were tipping the balance toward the resurrectionists, a group of Polish researchers in the early 1950s reported that some shriveled tardigrades were taking up minute amounts of oxygen from the air. Metabolism, they concluded, does not cease totally. In the 1960s, however, a curious paradox began to emerge. James S. Clegg, a young chemistry-oriented biologist then at the University of Miami, became intrigued by the brine shrimp "eggs" one of his colleagues was hatching to feed fish in a lab tank. The eggs, which are actually cysts containing partially developed embryos, are commonly sold in pet shops and can be stored dry in a bottle for years. Drop them in water, however, and within hours the dormant embryos resume development and eventually hatch into swimming larval crustaceans of the genus *Artemia*.

"When I started to think about it," Clegg recalls, "it was really amazing." Clegg put aside his research on the mechanisms of insect muscles and began to search for metabolism in brine shrimp eggs. He tested *Artemia* eggs that had been in storage for several years. Of one group that had been kept for fifteen years in air, less than 1 percent revived, but of another batch stored for ten years in a partial vacuum, 22 percent survived. Why, if cryptobiotic organisms metabolize oxygen, do they survive better when there is little or no oxygen to metabolize? There was an easy answer to this paradox, but finding it was not so easy. It turned out that oxygen consumption by the dried organisms was simply the nonmetabolic oxidation reactions that are no more a sign of life than is rusting. If oxygen is available, it can combine readily with some of the otherwise inert molecules that make up the dried animal, resulting in a deleterious buildup of terminally oxidized chemical compounds. These "adventitious" chemical reactions ought not to be confused with metabolism, Clegg observes, "otherwise one is compelled to conclude that such things as potato chips and Egyptian mummies metabolize." Turning rancid is not living.

The longer a dry organism is exposed to oxygen, the poorer its chances of reviving. Nonetheless, some cryptobiotic organisms are capable of surviving for startlingly long periods of time under ordinary conditions. The record appears to be held by certain rotifers and tardigrades that lay in a dried specimen of moss in a museum collection for 120 years. When moistened, some of the animals began moving around. Within a few minutes, however, they all died. Apparently the buildup of oxidation damage couldn't keep the simplest of life's motions from reviving, but it was not long before one or another vital metabolic process called upon molecules that had been hopelessly jammed by oxidation and the mechanical works of life simply stopped.

John H. Crowe, a zoologist at the University of California at Davis, is one of the few other modern researchers to work on cryptobiosis. He has found that for an animal successfully to enter the dessicated limbo in a condition that will permit it to emerge, the animal must dry slowly. A nematode pulled from damp soil and thrust into the sun will die within minutes, permanently.

"This is what confused a lot of the early experiments," Crowe says. "People would dry out some nematodes and try to revive them and fail. They doubted the phenomenon even existed. It turns out that the cryptobiotic state takes a little preparation. The animal has to get ready for it."

As the animal starts to dry out, it begins producing a compound not present in the active state—trehalose, a kind of sugar that combines two glucose molecules. If an animal dries out before it has manufactured enough trehalose, it will be unable to revive. When dried slowly, the animal shrinks, but no vital structures crack or separate. Instead, the organism's organs appear to fold and curl in an orderly way. The "skin" of a nematode, for example, pleats itself into an accordion fold. The entire nematode rolls into a snail-like spiral. Tardigrades contract and curl themselves into barrel shapes.

It appears that trehalose replaces water on the surfaces of all the structures of a drying animal, including all the organelles inside the cells. This would appear to prevent the animals from coming apart. The sugarlike compound also prevents bonding between proteins and ordinary sugars that may touch as cells shrink. Random bonding among various compounds can form a complex tangle of molecules that is insoluble in water. Such "browning reactions," as they are known, are common in dehydrated foods and help explain why these products seldom retain their original consistency or flavor when they are reconstituted.

So long as the structural integrity of the cells and the structural integrity of the tissues remain intact, the organism is alive, at least in the most fundamental sense. If enough protective trehalose can be synthesized before the loss of water stops metabolism, the animal's complete structural apparatus is preserved. As the dry organism absorbs water, tissues slowly swell. Trehalose is displaced from its protective positions on various surfaces inside the cells and goes back into solution. As a carbohydrate, it is readily metabolized for energy. All the proteins and nucleic acids reassume their active-state three-dimensional configuration and begin to react the way their chemistry dictates. In a sense, the organism springs back to life.

However extraordinary cryptobiosis seems to be, its death-defying power is limited. Nematodes and tardigrades and rotifers do eventually die. Though they may go in and out of the cryptobiotic state numerous times, the organisms do grow old and there comes a day when they die. For real. Like cats and dogs and people, their bodies grow tired and worn. Their muscles lose tone. Their tissues deteriorate. Their organs begin to fail. Eventually, as it apparently must to all multicelled organisms, death comes. The cells of cryptobiotic organisms lose their youthful ability to maintain the integrity of the structures within their cells and among their cells.

But why? Why do we have to die? Why do we even have to grow old? If many of the single-celled creatures of the world can keep the structures of life intact and working indefinitely and if structural integrity is what we need to keep us alive, why can't our cells do the same thing? Presumably if all our cells remained in as good condition as when we were young, our whole body would stay young and we, like the immortal protozoans, would never have to die. Is there some biological inevitability that catches up with us? Is it because of the most obvious difference between one-celled and many-celled organisms? Is there something about the republic of cells that demands that individual cells sacrifice their immortality?

There is considerable evidence that multicellular organisms are fated to die at a roughly predictable time. Mice usually die before they are 2 or 3 years old. Elephants rarely make it past 50 or 60. Most humans are dead before the age of 80 or 90 and virtually all by 120. The Galapagos tortoise appears to make it to 150 or more with relative ease but never much beyond 175.

At least since Aristotle, observations like these have led people to suppose that aging and death are somehow built into the fundamental makeup of each species. (The existence of single-celled creatures was not known, let alone their immortality.) Just as progeny always resemble parents in appearance, so they resemble them in life span. Long before anybody knew of genes, it was obvious that traits were passed from one generation to the next or, at least, that some mysterious force kept the traits consistent within a species, including the predictable inevitability of aging and death.

One great question remained, however: Do we age and die because our bodies are built to die, our deaths perhaps being necessary for the good of the species? Or do we die simply because our bodies break down after prolonged use, the species being indifferent to our deaths? In other words, is death biologically required or merely optional? The answer to that question could give us some hope that if we tried hard to stay healthy, we could live forever. Or it could assure us of personal doom.

In the 1920s a clue to the answer seemed to emerge from the experiments of Alexis Carrel, a French biologist who pioneered a number of early techniques of cell culture. Carrel, who worked at the Rockefeller Institute in New York City (now called Rockefeller University), established a line of chicken cells that he said was able to keep proliferating endlessly, generation after generation. The cell line was maintained for many years, outliving many times over the chicken that donated the cells. On the basis of this finding Carrel asserted that cells were not programmed to die. They were immortal and forever youthful, renewed with each division exactly as is the case with one-celled organisms. Whatever it was that made the whole chicken die—and by extension, what made any multicellular organism die—Carrel said, was that so many of the cells in the organism had become terminally differentiated. They had changed from their embryonic form, in which endless divisions were possible, into specialized forms—such as brain cells, muscle cells, heart cells, and so on—that could no longer divide.

We die, Carrel said, because our bodies are made up of so many specialized cells that can no longer undergo the rejuvenating influence of cell division enjoyed by one-celled organisms, even if those organisms are former members of multicellular bodies. Death, Carrel said, was the result of wear and tear within those undividing cells.

In the 1960s that seemingly reasonable view came crashing down with one of the most puzzling of all discoveries in cell biology. Leonard Hayflick, a cell biologist now at the University of California at San Francisco, found strong evidence that Carrel was wrong. Hayflick noticed that cell lines derived from normal (as opposed to cancerous) cells were not immortal. Over years of carefully controlled experiments, he showed that normal human cells can be grown in the laboratory for only about forty to sixty rounds of cell division. The average was around fifty. Then, even though the cells looked perfectly healthy, they lost the ability to divide any more and eventually died. The only cells Hayflick found that seemed to keep dividing endlessly were those from tumors. Cancer cells, as we have seen before, do appear to be "immortal."

What about Alexis Carrel's immortal chicken cells? Hayflick could not confirm Carrel's result. His own cultured chicken cell lines invariably became extinct after fifteen to thirty-five rounds of cell division. It is now generally accepted that Carrel's cell line was not really continuous. His method of feeding his cells used serum prepared from chicken blood in a way that inadvertently included fresh chicken cells in the preparation. The old cells, it now seems obvious, must have been dying off but, unbeknownst to Carrel, were being repeatedly replaced with new cells.

Hayflick's findings are now so well accepted that biologists who work with normal (noncancer) cells take care not to make the cells keep dividing too many times before they reach what has been called "the Hayflick limit." Instead of keeping the entire culture growing continuously, scientists freeze some of the cells early in their progress through their allotted number of divisions. Sibling cells may be kept alive and allowed to multiply until their number is up, but researchers can always go back to the frozen counterparts, thaw them out, and have the same cell line resume its dividing at an earlier number.

More profoundly, Hayflick's finding strongly suggested that the inevitability of death, and even a cellular form of aging, was built into each individual cell—or, at least, built into cells taken from multicellular animals. Microbial cells are still immortal. No matter how well cared for the human cells are, if they are not cancer cells, they invariably die after an average of 50 divisions. Death seems programmed into our genes.

Even more compelling evidence emerged when Hayflick made the same observations on cells of other species. Mice, which live at most for 3½ years, cannot keep their cells going in culture for more than fourteen to twenty-eight divisions. The Galapagos tortoise, which can live up to about 175 years, has cells that keep dividing for 90 to 120 rounds. In other words, the life span of a species appears related to the number of cell divisions its cells can sustain in culture.

Hayflick's findings could be regarded as among the most discouraging in all of science. They clearly imply that no matter how well we take care of our bodies, we cannot extend our life span beyond a fixed limit. But that implication need not be as dismaying as it seems. It is not clear that we die because our cells reach the Hayflick limit. The evidence is in cells removed from old people for culturing. The human cells that live for about fifty cell divisions are fetal or newborn cells. (Umbilical cords and foreskins are favorite sources.) Cells taken from people in their eighties or nineties typically live in culture for about twenty rounds of division. Most of us, it seems, die long before our cells reach the Hayflick limit. The implication of this finding is more hopeful: If we could prevent the things that kill us sooner, our cells may have the biological potential to carry us into our mid-100s.

To this potential good news, however, a discouraging caveat must be added. Even in a perfectly healthy body, some of the cells—including some of those most the crucial to survival—can no longer enjoy the revitalizing benefit of dividing. They dropped out of the numbers game early in development, committing themselves to specialized functions that make it architecturally impossible to divide again. This includes most muscle cells (which fused into giant supercells early in development) and, even more critical, nerve cells. While brain cells can change their wiring patterns, pulling back tentacles that made contact with one cell and sending them off to touch another, this ability to change shape is limited. For one thing, much of what the brain does when it learns is to create and maintain specific wiring patterns among brain cells. These wiring patterns are one of the ways skills are retained and memories are stored. If a nerve cell divided, it might renew itself as two young daughter cells but at the price of severing the connections needed to remember a fact or a skill. What's worse, since each brain cell may be in communication with tens of thousands of other brain cells, if it divided, it would cause a huge disruption. Brain cells that could keep dividing throughout adulthood might keep our brains from aging, but they might be able to store memories and skills for only a short time.

If, as anthropologists say, the great evolutionary adaptation of the human species is a superior brain capable of learning vast amounts of information, that specialization was bought at the price of an ability to repair the effects of damage and aging on the brain. Nerve cells have evolved to maintain their continuity to keep a lifetime's learning. Natural selection appears to have favored brains committed to storing memories over those whose cells freely divided. (Not that humans are the only species whose brain cells gave up the ability to divide. Learning and memory are essential to many species, and the sacrifice was undoubtedly made long before humans inherited that state of affairs.)

Still, it is not clear why a cell that has ceased dividing needs to die. Whether it stopped dividing within the body or reached its Hayflick limit in a laboratory flask, nobody really knows what happens inside a cell to bring death. Why couldn't a cell just keep on metabolizing forever? Is a cell like a machine whose parts wear out? Or do cells have the ability to replace worn-out parts with newly made ones? Or could it be both: Could it be that cells can replace some broken parts but certain irreparable parts are, nonetheless, vital to life?

At least a dozen hypotheses, from the fanciful to the plausible, have been advanced to explain how a cell might lose its capacity to keep on living.

One possibility is that after a cell becomes fully specialized—or what the jargon calls "terminally differentiated"—it has so much specialized work to do that there is simply not enough metabolic capacity to keep making the enzymes and structures needed for the perfect housekeeping that would keep a cell young. So the original stockpile of vital parts is eventually damaged beyond the ability to perform.

Another idea is that mutations are hitting the cell's genes from time to time and that as the amount of damage grows, the odds get higher that vital genes are knocked out. A liver cell, for example, might lose the ability to make an enzyme needed to carry out some vital function for the body. As more such cells are damaged, the liver's efficiency declines with age. Without a minimally functioning liver, the rest of the body is doomed.

Another school of thought blames the immune system for the deterioration of the aging body. The idea is that the immune system gradually loses control and begins mistaking parts of its own body for enemies. There is good evidence that as people get older, their blood contains more and more of what are called autoantibodies, antibodies shaped to bind to molecules and cells of one's self. As yet, however, there is no clear link between the types of autoantibodies known to arise and the diseases that are most common among the elderly.

Yet another hypothesis is called "error catastrophe." The idea involves a healthy cell making a tiny error when translating the genetic code for a protein needed in the very process of translating genes into proteins. Since this would produce defective translating machinery, that new machinery would do a poorer job of translating the next generation of proteins. Conceivably every protein made by the flawed translating process would itself be flawed. This rather clever idea suggested that a tiny initial error would quickly mushroom into a catastrophe, filling up the cell with enzymes that couldn't carry out the right reaction and structural proteins that wouldn't assemble into proper structures.

Another theory for aging of cells invokes the well-known fact that when various proteins are subjected to a certain amount of heat, they become permanently deformed. The most familiar such reaction is what happens to egg white, which is largely the protein albumin, in a frying pan. Living cells don't normally get that hot, but gentler heat can produce gentler gumming up of proteins. Some proteins, like albumin, become cross-linked as bonds form between molecules that are supposed to stay separate. Other proteins when heated, "melt" into unusable shapes and can't be put right again. A cell with too many of these so-called denatured proteins would be in a bad way.

A related hypothesis blames the way glucose molecules can cross-link nearby proteins, taking them out of service like two shoes with laces mischeviously tied together. This cross-linking is one of the effects that would kill cryptobiotic animals when they dry up were it not for the protective effects of trehalose.

Still another suggestion involves the effect of ordinary water on certain molecules—a chemical reaction called hydrolysis. What happens is that water breaks apart certain molecules essential to life, including proteins and, under certain circumstances, even DNA. Hydrolysis splits them into two fragments and attaches the hydroxyl group (the oxygen atom and one of the two hydrogens) to one piece and the leftover hydrogen atom to the other. While hydrolysis plays an essential role in many metabolic reactions, it could obviously be rather disruptive where it was not wanted—like breaking apart a gene. Fortunately, since water is abundant in cells, DNA is actually rather resistant to hydrolysis under most conditions, and even if it happened, there are DNA repair enzymes that can spot the problem, remove the intruding oxygen and hydrogens, and splice the DNA back together. Still the repair mechanism could conceivably become faulty or simply fail to catch a break.

Yet another theory of aging and death invokes a curious phenomenon in which an amino acid suddenly flips itself inside out, something like an umbrella caught in a gust of wind. This does happen. Amino acids, like many other molecules, are not rigid. There is flexibility at the joints where one atom is linked to another, and many of life's molecules are constantly flexing and jiggling, changing their shapes ever so slightly. Sometimes the physical forces that account for this motion will randomly combine to make the atoms move past one another more than usual. Then the molecule suddenly flips into a different but stable arrangement. As with the umbrella, the new shape may be functionally useless. If the amino acid happened to be in a vital protein when it flipped inside out, the protein could be rendered useless.

All of these theories have had their advocates and their enemies. None is fully accepted, although there is room to think that any one of them, or several, could be playing small roles in the aging and death of cells. Three additional hypotheses, however, have gained wider followings.

One is the cellular equivalent of indigestion, or perhaps it could be called terminal constipation. From time to time cells take in substances that they cannot fully digest in their lysosomes. This indigestible junk accumulates in the cell. Single-celled organisms eventually excrete this stuff, disgorging it in the cellular equivalent of defecation. One of the great sacrifices made by many of the cells that have banded into multicellular organisms, however, is that they have given up the ability to do this. And for good reason. The lysosomes contain caustic digestive enzymes that if released from the protection of the lysosome's membrane, can damage cells and the extracellular matrix. There seems to be no way for cells to unload their lysosomes without exposing their neighbors to these dangerous enzymes. As long as cells can divide and grow back to full size, they can dilute the amount of accumulated junk. But terminally differentiated cells lose this option. More and more useless stuff piles up in these cells.

Consider the nerve cells, which have not divided since childhood and may be many decades old. Most of the cell will be young because its internal renewal processes—breaking down old organelles and enzymes in the lysosomes and assembling new ones—will have recycled most of the organelles. Even in a seventy-year-old brain cell, most of the organelles will be no more than a few weeks old. Over the years the cell will have broken down (in the lysosome) and remade most of its molecules and structures thousands of times. But there is one exception to this process of continuing cellular renewal—the brown, indigestible debris that accumulates in the lysosomes. Cell biologists call the stuff lipofuscin.

The lipofuscin theory of cell aging is that as the gunk accumulates, cells become less and less able to continue the processes of renewing all their other molecules and structures. Lysosomes are essential to that process, for they break down the old hardware of life, reducing it to its raw materials, which are then recycled for new construction. A lysosome jammed with lipofuscin would not seem able to continue that process efficiently. Cells would gradually lose their ability to perform their jobs, and the republic would gradually disintegrate.

Lysosome troubles, incidentally, are also at the bottom of some of the degenerative diseases of aging. Rheumatoid arthritis, for example, is the result of cells of the immune system, such as macrophages, trying to engulf huge objects such as the extracellular linings of joints that have become covered with misguided antibodies. As the macrophage spreads itself over the surface, in a desperate attempt to swallow something thousands of times larger than itself, its lysosomes rupture. The digestive enzymes leak out and chew away some of the joint lining.

The other theory of aging and death that has earned a number of adherents involves entities called free radicals. That the term evokes notions of political revolutionaries bent on destroying the established order is inadvertent. And yet atoms and molecules that are free radicals can seriously undermine the structural integrity of a cell. In chemistry a radical is an atom or a molecule (a group of bound atoms) that has a high chemical propensity to bind, as a unit, to another atom or molecule. (The term comes from the Latin *radicalis* for "root," as in the root vegetable radish. Chemical radicals are often the root components of larger molecules. When the radical is freed from the rest of its larger molecule, it is a free radical.)

Free radicals are produced inside cells when certain compounds react with oxygen and when radiation, including light, strikes certain molecules, breaking them into fragments. Heat can have a similar effect, as can certain toxins. Whatever the cause, the result is an atom or a molecule with an unpaired electron. Since an atom's electrons must exist in balanced pairs for the molecule to be stable, such entities are highly reactive and will stop at nothing (almost) to get access to another electron. Most free radicals have a powerful propensity to bind to some other atom or molecule, even if it was happily stable, or even to break apart another molecule to steal an electron.

This would not be so bad if the process stopped there. A healthy cell could easily compensate for the loss of one molecule every now and then. But what often happens is a chain reaction. As the first free radical rips apart some law-abiding molecule to satisfy its chemical drives, it leaves fragments that themselves become free radicals. Like Dracula's victims, they must now go out and find a blood meal for themselves. One of the best-known effects of free radicals involves fatty molecules. Often these are transformed into toxic hydrocarbons such as ethane and pentane. A chain reaction that went unstopped could easily destroy the life-sustaining, antientropic structures within a cell and stuff it with uselessly mangled molecules. Many of these cannot be digested by the lysosomal enzymes, so they accumulate as part of the lipofuscin.

Fortunately, cells have ways of protecting themselves from free radical damage. One is a special enzyme that unradicalizes the free radicals, sometimes by causing two free radicals to bind with each other. As long as the cell is well supplied with this enzyme, called superoxide dismutase, it is relatively well protected against free radical damage. But, as with all the protective mechanisms of life, the mechanism can fail. Or the rate of free radical chain reactions can simply outrun the rate of the protective reactions. Even if the protections catch up, however, too many toxic chemicals may already have been produced.

Ironically, one of the most potent free radicals is oxygen. It reacts vigorously with a wide variety of atoms and molecules, from iron (to make rust) to hemoglobin (from which it, fortunately, is liberated to supply cells) to the reactions within cells where oxygen is used to "burn" carbohydrate molecules to release their stored energy. The cellular enzymes that deal with oxygen are, of course, usually very good about keeping the radical under control. But yet again, there can be slip-ups where the occasional oxygen will escape and wreak molecular havoc.

Biochemists have long known that it is possible to quench the electron thirst of a free radical with another kind of molecule called an antioxidant. These are molecules that carry a surplus electron that can be given to a free radical (be it oxygen or any other) without themselves turning into something harmful to the cell. This has led to the speculation that if cells were well supplied with antioxidants, they would be better protected against the damage inflicted by free radicals.

Intriguingly, vitamins C and E are known to be antioxidants. Vitamin E is used as a food additive to retard rancidity, which is simply the oxidation of food molecules (the same process that can cause dormant cryptobiotic organisms to appear to be consuming oxygen). Researchers have conducted a variety of experiments to see whether these vitamins combat aging. One study has found that people with vitamin E deficiency have cells with increased amounts of lipofuscin. Another, even more tantalizing, found that mice fed extra vitamin E lived longer than mice on an ordinary diet. There is one caveat in this study, however. The mice also lost weight on the diet and it is known that low-calorie diets— nutritionally balanced but just barely meeting the needs of the animal— also extend lifespans.

While studies like these have led some to prescribe vitamin supplements for people, it is already clear that antioxidants cannot stop aging and cell death altogether. Cell cultures fed extra vitamin E still died out when they reached the usual Hayflick limit—roughly fifty cell cycles for human cells. But, of course, since it appears most people die decades before they reach the Hayflick limit, antioxidant supplements might still be useful in maintaining the structural integrity of cells up until that limit is reached.

The newest hypothesis of aging, and one of the most exciting, involves telomeres, the monotonous strings of genetic coding found at the tips of all chromosomes. These are the repeats of the sequence TTAGGG, which as we saw in Chapter 12, are lost at each round of cell division except in sex cells. This is an unavoidable consequence of the way DNA polymerase works; it simply is unable to copy the tips of chromosomes. This is usually not a problem because the telomeres are there precisely to be lost so that the genes that lie toward the centers of chromosomes are protected.

But, of course, since a few score telomeric bases are lost at each round of cell division, the time will come in a frequently dividing tissue when the next round of cell division will cut into the bases of a gene needed in that cell. The daughter cells will be deprived of a resource they need to carry out their function in the body, and one organ or another will begin to malfunction. The cell may also be deranged by the loss of another role telomeres play. Telomeres stabilize chromosomes. Without them, chromosomes can link together or break into segments that rejoin in abnormal combinations. These alterations could also cripple cells, causing them to senesce and, as they deprive the body of needed functions, leading to bodily aging.

Although the accumulation of lipofuscin, the related effects of free radical damage and the loss of telomeres are perhaps the leading hypotheses to account for aging and death, there is certainly no broad consensus among scientists. Quite possibly we age and die because of a combination of processes including these and some of the other phenomena mentioned earlier. There may also be causes as yet unknown. Living cells are so extraordinarily complex—thousands of different chemical processes being needed to maintain the structures that make life possible and to exercise the various motile powers of life—that there would seem to be any number of things that could go wrong to kill a cell, or to cause the cell to fail in its service to the republic, or to cause it to abandon some vital function upon which other cells depend.

And yet we know that free-living, single-celled organisms need never die. They are as complex as any of our cells, yet they are able to repair the ravages of time. Cell division for them would seem to be a fountain of youth. Pass through it and you never die. When single cells banded together hundreds of millions of years ago to form multicelled organisms, each cell had to subordinate its freedom to the good of the community, to the good of the republic. Each cell also surrendered its immortality. Refraining from the dumping of lysosomal wastes was one evidence of those sacrifices. Its price may have been the gradual destruction of any given cell's ability to renew its parts indefinitely.

Even if that were found to be the root cause of cellular aging and death (and hence the death of the whole organism), there would remain a mystery. Why do mice die after two or three years while humans live many decades longer. Are our cells that different biochemically? They must be.

To Leonard Hayflick this suggests a different approach to the riddle of mortality. The question, he says, is not "why do we age?" but "why do we live as long as we do?"

The principles of evolutionary biology suggest an answer. To ensure the survival of the species, individuals need to live only long enough to see that their offspring are well launched on their own lives. Consider the salmon, whose young need no help from their parents. Once the salmon swim upstream and spawn, they die. The salmon species has no further need of those particular individuals. The spark of salmon life has been passed in egg and sperm to a new generation.

Indeed, Alfred Russel Wallace, the co-discoverer with Darwin of natural selection, once argued that parents that lived too long would eventually compete with their young for resources. Survival of the young, once they are properly reared, may thus be improved if the old make the sacrifice of dying off. Wallace thought the time of death might be a trait that evolution programmed into the species as if it were a pattern of coloration or a body shape. If he is right, the Hayflick limit may be the evidence that the time of death is programmed. It comes very late under the ideal conditions of cell culture in the laboratory. Perhaps it comes sooner under the natural conditions of the body because of various adverse effects inflicted by other cells living in close proximity. Since the timing of the cell cycle is different in different tissues, it may be that some cells in vital organs are on a fast track and reach their Hayflick limit before cells in other tissues. But those other tissues will be doomed nonetheless when the fast-track cells give out.

Mammals, especially the more social ones, need longer parental care than fish. The newborn mouse must be suckled for a few weeks and then trained in the ways of mouse society. Parent mice must be strong enough and healthy enough to live through this period. To do so, they must have cellular mechanisms that can repair the damaging effects of age for at least that long. But once the young are launched—a matter of weeks in the mouse—the species no longer needs the parents. Mice typically die after only about two years. There is no evolutionary pressure for mice to live any longer. Natural selection can favor a longer lifespan only if that result somehow affects the survival of the next generation carrying those genes.

Humans, requiring a very long period of parental care, could not have evolved if their cells aged and died as quickly as those of mice. Even if one remembers that virtually all of human evolution took place when our species lived as small bands of hunters and gatherers, a child would still benefit from having living parents until perhaps adolescence, when the child would be ready to have children of his or her own. This means survival of the human species would require that people live at least until, say, the age of thirty. Since at least two children must reach reproductive age for each couple, this might well mean the parental generation should be biologically fit until perhaps forty years of age. Why, then, do humans live well past this age? There might also be a premium on living still longer if the wisdom and knowledge of the elderly benefited the survival of the social group. The respect accorded old people in many traditional human societies may be a relict sign that our species evolved by depending on the wisdom of the elderly.

Hayflick suggests that evolution has favored humans whose cells can keep their internal structures intact long enough to launch the next generation and that if we live longer than that, we are simply coasting, like a race car that has crossed the finish line, on the reserve capacity built into the system. In other words, evolution has guaranteed us cells that can live in a multicellular organism perfectly well until we reach what is now known as middle age. Then our cells slowly start falling apart. The structural integrity that makes for life—whether in the dessicated cryptobiotic nematode or in the geriatric human—does slowly succumb to the law of entropy. Cells fall apart in the course of living. But they can be repaired as good as new as long as they can keep building new structures to replace the old, as long as they can keep capturing new nonliving matter and organizing it into the correct patterns and shapes. And they can keep doing this as long as free radical damage has not crippled vital enzymes and structures and as long as the lipofuscin buildup has not clogged their lysosomes' abilities to help recycle the materials needed to rebuild.

When death comes to most human beings, however, it is not usually complete death. Not all of our cells die. A very select few of our cells, perhaps not more than one or two or three, do often find immortality. These are, of course, the sex cells—the sperm and eggs that have joined to begin constructing a new generation of human being. These cells possess the spark of life received from our parents and are carried forward to our children. The sex cells relay life to swarms of new cells that will build another human body. And, eventually, a few of the cells from that next generation will pass the torch to yet another.

The immortality of the single-celled organism is in us. The grand architectures of our sixty-trillion-celled bodies are but contrivances of the sex cells bent on sustaining their immortality. From an evolutionary perspective our bodies are merely machines whose purpose is to convey the sex cells into their selfish futures. As generations of biologists have noted, "A chicken is an egg's way of making another egg."

The republic of cells, it turns out, is ruled by an immortal autocrat— the molecular architecture that is the essential core of life, the structural integrity that must be preserved and transmitted to a new generation. In a sense, the parts of us that are not sex cells are expendable hardware, throwaway vehicles whose job is done when our precious cargo is safely transmitted to a new generation.

And yet we human life-forms are something more. The bacterium and the protozoan serve the goal of mere immortality as well as we. The vast assemblage of cells that nurtures the spark in each of us—though at the price of personal death—has given rise to the far more profound manifestations of consciousness and of curiosity and of reason. Through our science we are making life known to itself.

# GLOSSARY ●

**actin**  One of the two main kinds of protein filaments that make up muscle and the parts of cells that generate movement. Myosin molecules pull on actin to generate motion. Actin filaments are polymers of subunits also called actin.

**ADP**  Adenosine diphosphate. An ATP molecule from which one phosphate has been removed, liberating the energy in its bond for use by the cell. See **ATP**.

**animism**  An eighteenth-century theory that all components of the body were passive and that motion, or animation, was a result of the soul acting on the body.

**antibody**  A type of protein made by the immune system's B cells. Antibodies are shaped to bind to one specific kind of molecule, often called an antigen. Antibodies that bind to disease-causing organisms help kill them.

**antigen**  Any molecule, though usually a protein, that causes the immune system to make antibodies shaped to bind to it. These are typically foreign molecules and disease organisms.

**antioxidant**  Any molecule that blocks or reverses oxidation, thus preventing the creation of cell-damaging free radicals.

**apoptosis**  Programmed cell death, a phenomenon in which cells commit suicide when deprived of sustaining signals from neighboring cells.

**ATP**  The universal energy-carrying molecule of cells. It is made in mitochondria with the energy obtained from food and sent to all other parts of the cell. Energy is released when one phosphate is removed, leaving ADP.

**axon**  The long, filamentous projection a nerve cell sends out to connect with distant nerve cells.

**base**  The portion of a DNA strand that projects from the "backbone." When considered with its adjacent segment of backbone, it is called a nucleotide. There are four kinds, they can occur in any linear order, and their sequence encodes the gene's message.

**blastocyst**  An early stage of development in which the embryo (or, more properly, the pre-embryo), consists of a hollow ball of cells with an "inner cell mass."

**carcinogen**  A substance that causes cancer.

**cdc protein**  Any of a class of "cell-division-cycle" proteins needed to regulate the cell's life cycle.

**cell**  The smallest unit of matter that can be considered alive. The fundamental particle of life.

**cell cycle**  The sequence of events that cells go through to divide, perform their normal duties, and divide again. Beginning at M, the cycle proceeds to G1, S, and G2 before returning to M.

**cell division**  The highly orchestrated process by which one cell duplicates its chromosomes, apportions them to two nuclei, and splits to form two cells, each with one nucleus.

**cell line**  A population of cells of a given type grown in the laboratory.

**cell membrane**  The outer "skin" of a cell, also called the plasma membrane. It is made of two layers of phospholipid molecules held together by intrinsic hydrophobic and hydrophilic properties.

**cell memory**  As embryos develop, their cells become progressively more specialized in shape and function. With each round of cell division, the new cells "remember" the specializations acquired by their "parents."

**cell theory**  Advanced in the nineteenth century, it says organisms are made up of smaller units of life called cells and that new cells are formed from the division of old cells.

**centriole**  A cylindrical array of short microtubules. Two centrioles form the center of the cell's centrosome. During cell division, they separate and produce the microtubule fibers that will help separate the two sets of new chromosomes.

**centromere**  The place on a condensed chromosome (a form taken to prepare for cell division) where duplicated chromosomes are held together temporarily. This is also where structures called kinetochores will form to grasp microtubules from the centrioles to separate the pairs.

**centrosome**  The structure within cells from which a vast network of microtubules radiates. It contains two centrioles. Also called the "cell center" and "microtubule organizing center."

**channel**  A hole in the cell's outer membrane lined with proteins that control the movement of specific kinds of molecules into or out of the cell.

**chaperone**  A protein that assists a newly forming chain of amino acids to fold into the proper shape to function as a protein.

**cholesterol**  A kind of fatty molecule that is an essential component of cell membranes but that can also form concentrations on artery walls, potentially blocking them.

**chromosome**  A very long strand made of the chemical called DNA with its associated proteins. It is made up of four kinds of units called nucleotides, whose sequence encodes all the hereditary information of a cell. Traditionally, the word refers to the condensed form that the DNA molecule takes on for cell division. Nowadays, it also refers to the stretched out form of the same molecule.

**cilia**  The plural of *cilium,* a short, hairlike projection from a cell that contains microtubules and can beat or wave.

**clathrin**  A protein that links up with others of its kind to form a cagelike structure that forces the cell's outer membrane to form a pit into the cell. The molecules ultimately cause the pit to close into a bubblelike vesicle containing substances that the cell takes in.

**cleavage**  The act of a dividing cell pinching in two.

**codon**  A group of three nucleotides (also thought of as the three bases that are part of those nucleotides) that specify one of the twenty kinds of amino acids to be linked to make a protein.

**collagen**  A fibrous protein secreted by cells and used to make parts of the "extracellular matrix" on which cells crawl. Collagen is also an important component of cartilage, tendon, and bone.

**compaction**  A step in early embryonic development in which each cell in a loose ball produces junctions that pull it tight against its neighbor cells.

**complement**  Proteins produced by cells of the immune system that can cooperate to kill invading microbes and encourage immune system cells to eat the microbes.

**cryptobiosis**  A form of suspended animation (literally, "hidden life") into which certain cells and small organisms can enter when they lose most of their water. Cryptobiotic organisms do not perform metabolic processes, but may resume doing so after months or years when rewetted.

**cyclin**  One of the proteins that govern the cell cycle. Its concentration rises and falls periodically, activating other proteins in turn. The entire process is like a chemical clock that tells cells when to perform the steps needed to divide.

**cytokine**  A protein released by cells to signal other cells to take some action or initiate some process.

**cytoplasm**  A general term for all of the components of a cell other than the nucleus.

**cytoskeleton**  A catchall term for various kinds of filamentous networks inside cells. They include actin, microtubules and so-called intermediate filaments. They give the cell shape and the capacity to crawl and also serve as tracks for internal transport of vesicles and other organelles.

**daughter cell**   Either of the two new cells born through cell division.

**desmosome**   A type of strong junction that binds adjacent cells.

**DNA**   The name of the chemical of which chromosomes are made. It is a chain of subunits that repeat indefinitely. See **chromosome; gene.**

**DNA polymerase**   The large, complex enzyme that makes copies of DNA strands in preparation for cell division.

**docking protein**   A molecule for which a docking site exists. See **docking site.**

**docking site**   A place on a protein molecule (or complex of several proteins) where another molecule may fit with near perfect conformity. Generally synonymous with **receptor.**

**double helix**   The name of the shape of DNA when it is in the usual, double-stranded form. The rails of a spiral staircase form a double helix.

**dynein**   One of several motor molecules that transport objects about inside cells. Dynein moves in a direction opposite to that of kinesin.

**embryonic disc**   The double layer of cells that forms inside the blastocyst and in which the embryo will begin to form.

**endocytosis**   The act of a cell "swallowing" materials from outside. The outer membrane dimples into the cell (with the aid of clathrin) and closes over a small volume of the outside fluid. This forms a bubble of membrane, a vesicle called an endosome.

**endoplasmic reticulum**   A vast, convoluted structure where newly made proteins are chemically modified. Also called the ER, it is enclosed by a membrane.

**endosome**   A membrane-bound organelle, or vesicle, that transports materials ingested by the cell (via endocytosis) to a lysosome for digestion.

**enhancer**   A segment of DNA near a gene to which a regulatory protein may bind to help switch on the transcription of a gene.

**enzyme**   Any protein whose job is to facilitate a chemical process. Enzymes make chemical reactions happen (or happen faster) but are not consumed in the process.

**epithelium**   A sheet of tightly joined cells that form a relatively impervious covering of a tissue or lining a cavity.

**ER**   Endoplasmic reticulum.

**exon**   A nongibberish part of a gene. It often encodes a functional module of the intended protein. The other parts are called introns.

**extracellular matrix**   Proteins (such as collagen and fibronectin) made by cells and released outside (secreted) to form surfaces or scaffolding of tissues.

**fertilization** The fusion of sperm and egg, a process that includes several steps.

**fibroblast** A type of cell that secretes extracellular matrix such as collagen and fibronectin. Also a favorite type of cell for culturing.

**flagellum** The singular of *flagella*. A long, hairlike projection from a cell, filled with microtubules. Flagella are essentially long cilia used as propulsion mechanisms by free-swimming cells.

**free radical** Molecules that have an unpaired electron (often a result of oxidation), which makes them react strongly with other molecules.

**G1 phase** Gap 1, the interval in the cell cycle during which cells perform their normal duties.

**G2 phase** Gap 2, the interval in the cell cycle after chromosomes have been duplicated and before M, or mitosis, when the duplicate chromosomes are separated.

**ganglion** The singular of *ganglia*. A cluster of nerve cells outside the central nervous system.

**gap junction** Small portals in adjacent cells' outer membranes through which small molecules pass freely from one cell to its neighbor.

**gene** A length of DNA that contains the complete code for guiding the synthesis of one protein or of an RNA molecule. The term is often used to include the noncoding "introns" and associated regulatory sequences.

**genetic code** The set of rules that determine which amino acid is specified by three adjacent nucleotides of DNA, called a codon. The phrase is sometimes used to mean a specific set of instructions for making one protein.

**Golgi apparatus** A membrane-bounded organelle in which proteins made in the ER are further processed and sorted.

**growth factor** A protein that stimulates cells to divide when it binds to an appropriate receptor on the outside.

**hemoglobin** A complex of four proteins (two alpha hemoglobins and two beta hemoglobins) that picks up oxygen in the lungs and releases it in other tissues. It is carried inside red blood cells.

**homeobox** A DNA sequence important in embryonic development. It encodes a protein that activates the expression of many different genes that must work together to give rise to a particular region of the body.

**hydrolysis** The splitting of a molecule in which a water molecule is also split, its H joining one product of the split and its OH joining the other.

**hydrophilic** "Water loving." Said of a molecule or region of a molecule that has an affinity for water. The opposite of hydrophobic.

**hydrophobic**  "Water hating." Said of a molecule or region of a molecule that is water repellent. The opposite of hydrophilic.

**immune system**  A variety of cells and tissues that combat infection. It includes white blood cells (lymphocytes), lymphatic vessels and lymph nodes, spleen and thymus.

**inner cell mass**  A clump of cells inside the otherwise hollow early embryonic stage called the blastocyst. The ICM will give rise to the embryo proper.

**intermediate filament**  One of three main types of cytoskeleton. Named because the diameter of the filaments is between that of the other two types of cytoskeleton. There are several kinds of intermediate filaments.

**intron**  Intervening segments of noncoding DNA in genes. The other parts are called exons.

**ion pump**  A structure in the outer membranes of cells that use energy to move ions (usually calcium, sodium or potassium) from one side to the other.

**kinesin**  A motor molecule that transports vesicles of cargo along microtubule tracks. It moves in a direction opposite to that of dynein.

**kinetochore**  A structure on a chromosome during cell division. Kinetochores on duplicated pairs grab spindle filaments (microtubules) reaching toward the chromosomes from opposite sides and pull the two chromosomes apart.

**lymphocyte**  A white blood cell that is part of the immune system. The two main classes are B lymphocytes and T lymphocytes.

**lysosome**  The cell's stomach. It contains digestive enzymes that break down large molecules taken in by the cell and worn-out organelles being recycled. A cell has many lysosomes.

**M phase**  The period of the cell cycle in which cells divide their nuclei and split apart. *M* is for mitosis.

**macrophage**  Literally, "big eater." A free-roaming cell that consumes dead and dying cells of the organism as well as invading microbes.

**mast cell**  A type of cell found in many tissues that releases histamines as part of the inflammatory response, a primitive form of immune response.

**mechanist**  A person who holds that all the processes of organisms happen as a result of entirely natural processes. This includes virtually all cell and molecular biologists today but some decades ago, there were also "vitalists," who believed that some supernatural force caused motion and other activities of living organisms.

**meiosis**   The process in ovaries and testes in which sperm and eggs are produced with only half-sets of chromosomes. It involves two rounds of cell division but only one duplication of the chromosomes.

**mesoderm**   The middle of the three founding layers of an embryo.

**messenger RNA**   A transcribed copy of the DNA code of a gene but in a slightly different form. It carries the genetic instructions for a specific gene from the nucleus to the ribosomes in the cytoplasm. Also called mRNA.

**metastasis**   The process by which cancer cells break away from a tumor and move to other parts of the body.

**microtubule**   One of three kinds of cytoskeleton. During cell division, microtubules act as spindle fibers, helping to separate duplicated chromosomes. At other times, they form tracks for transporting molecules throughout the cytoplasm. Each microtubule is a polymer of tubulin molecules.

**mitochondrion**   The organelle that packages the energy in food into ATP, the universal energy carrying molecule in cells. Every cell has many.

**mitosis**   The separating of duplicated pairs of chromosomes to form two identical nuclei.

**monocyte**   A type of white blood cell that can leave the blood stream and change into a macrophage.

**monomer**   A subunit, or building block, of a polymer.

**morula**   In humans, the sixteen-cell stage of embryonic development. Just before compaction, the cluster of cells resemble a mulberry (Latin, *morula*).

**motor molecule**   Any of several kinds of proteins that can use the energy in ATP to propel themselves along a filament or across another molecule.

**muscle fiber**   A muscle cell, it arises by the fusion of what once were many smaller cells.

**mutation**   A change in the nucleotide sequence, or DNA code, that leads to a change in the resulting protein.

**myosin**   One of the two main protein filaments of muscle cells. Myosins pull on actin filaments to make muscles contract. In other cells, myosin also pulls on an actin meshwork to change the cell's shape and cause it to crawl.

**natural killer cell**   A type of immune system cell that binds to cancer cells and cells infected by viruses and kills them.

**natural selection**   An inescapable result of the fact that some organisms and cells will die prematurely if not suited to their environment. The process thus naturally selects those that are suited to survive and reproduce.

**necrosis**  A pathological form of cell death, typically caused by injury or infection, in which the degradation products elicit an inflammatory response. Contrast with **apoptosis**.

**neural crest cell**  A type of cell that begins life on the early embryo's neural tube, but then migrates to other parts of the embryo to give rise to some nerve cells, skin pigment cells (melanocytes), and certain facial bones, including both jaws.

**neuron**  A nerve cell.

**neurotransmitter**  A substance emitted by one neuron to act on another neuron or on a muscle or gland. Some neurotransmitters move across the tiny gaps called synapses; others diffuse more widely to affect numerous surrounding neurons.

**nucleolus**  The organelle that make ribosomes.

**nucleotide**  One unit of a DNA strand, consisting of a segment of the "backbone" plus one of the four kinds of base.

**nucleus**  The largest organelle in a cell, it contains the chromosomes plus the nucleolus.

**oncogene**  Any of a large number of genes that, when damaged, acquire a function that turns a cell cancerous.

**operator**  A regulatory segment of bacterial DNA that controls the transcription of an adjacent gene.

**organelle**  "Little organ." An internal organ of a cell.

**phagocyte**  A cell specialized for eating, especially microbes and dying cells. Macrophages are a prime example.

**phagocytosis**  The process by which certain cells engulf and consume bacteria or other cells.

**phosphorylation**  The addition of a phosphate molecule to some other molecule. The effect is to so alter the shape of the receiving molecule that it either gains or loses functional ability. Thus phosphorylation is a simple way to turn processes on or off.

**plasma membrane**  The outer membrane enclosing a cell.

**polymer**  Any molecule made by linking smaller molecules, or monomers, into long chains. DNA is a polymer, as is protein.

**polymerize**  The process of monomers linking into chains.

**programmed cell death**  Also known as apoptosis, this is the process by which cells commit suicide when deprived of signals from their neighbors or when told to by special "killer" cells.

**promoter**  A segment of DNA adjacent to a gene and to which the enzyme DNA polymerase can bind to begin making an RNA copy.

**protein folding**  The process by which a chain of amino acids curls, flexes, and wads up into a three-dimensional form whose outer contours have specific abilities. The pattern of folding is dictated by the specific sequence of amino acids and the way in which they link to one another.

**protein kinase**   An enzyme that transfers a phosphate from ATP to a specific amino acid in a protein. This act of phosphorylation may change the protein into an active or inactive form.

**proto-oncogene**   The normal form of a gene that, when mutated, becomes a cancer-causing oncogene.

**protoplasm**   An archaic term for the stuff inside cells.

**pseudogene**   A sequence of nucleotides that resembles a known gene but that is not functional because it contains errors that prevent expression. Some may be deactivated forms of genes used in an evolutionary predecessor and some may simply be products of inadvertent gene duplication with errors.

**pseudopod, pseudopodium**   The "false foot" protruding out from a crawling cell.

**receptor**   A protein embedded in a membrane with a shape that can bind a free molecule with a complementary shape. In response, the receptor may initiate some process within the cell or may bring the free molecule into the cell.

**regulatory sequence**   Any of several kinds of DNA coding near a gene that help govern when the gene is expressed. The regulatory sequence does not encode any part of the resulting protein.

**ribosome**   The small organelle that reads the code in messenger RNA and follows its instruction to assemble a chain of amino acids that will become a protein.

**RNA polymerase**   The enzyme that reads the DNA code for a protein and assembles a molecule of messenger RNA carrying a transcription of the code.

**S phase**   The part of the cell cycle in which cells are duplicating (S for synthesizing) copies of their chromosomes.

**secrete**   To release molecules, usually by pouring them out through the plasma membrane.

**self-assembly**   The phenomenon by which certain kinds of molecules automatically link with one another to form larger structures.

**somite**   One of a series of domains within the early embryo where specific tissues form.

**spindle**   The apparatus that helps duplicated chromosomes separate during mitosis. It consists of centrioles at opposite poles of the cell that send out microtubules that attach to individual chromosomes. Early microscopists thought the array looked like a weaver's spindle.

**spindle fiber**   One of the microtubules reaching out from a centriole in a spindle during mitosis.

**synapse**   The place where one nerve cell releases its neurotransmitter chemicals to act on another.

**telomere**   A set of repeated nucleotides that serve as "dummy" DNA at the tips of all chromosomes. Because some nucleotides fail to be copied when chromosomes are duplicated for cell division, the telomere provides sequences that usually can safely be lost while protecting genes from being damaged.

**tight junction**   A type of linkage between adjacent epithelial cells that makes the sheet relatively impervious.

**transcription**   The process in which the DNA code is copied into the form of RNA.

**transfer RNA**   A small form of RNA that carries one amino acid. Each has three nucleotides that, under the control of a ribosome, will bind to a complementary sequence of one codon of messenger RNA.

**translation**   The process in which messenger RNA's code is read and the specified chain of amino acids is assembled.

**tubulin**   A subunit of a microtubule.

**tumor suppressor gene**   A class of genes whose normal function is to moderate cell division. So when the gene is mutated so that it no longer functions, a cell loses one of the restraints that kept it from becoming cancerous.

**vesicle**   Literally, "small vessel." A small, membrane-bounded structure that may enclose a cargo of molecules. Vesicles are tiny containers of cargo being transported within a cell.

**vital force**   A mythical, supernatural force that scholars of a century or more ago invented to explain extraordinary properties of life, such as the ability to move.

**vitalism**   The school of thought that invented the "vital force."

**zona pellucida**   A jellylike shell surrounding the ovum. Upon fertilization, it hardens, keeping other sperms out.

**zygote**   A fertilized ovum.

# Suggested Readings ●

These books will take readers further into the realms of molecular and cell biology. Most are textbooks, which, short of a visit to a laboratory, are the best way to learn more.

Alberts, Bruce, et al. *Molecular Biology of the Cell.* 3rd ed. New York: Garland, 1994.

This is the premier textbook of cell biology, technical and sophisticated but loaded with diagrams and explanatory graphics. For the serious student of cell biology, it is a must.

Darnell, James F., et al. *Molecular Cell Biology.* 3rd ed. New York: Scientific American Books, 1995.

This is the second most popular textbook of cell biology, covering much the same ground as *Molecular Biology of the Cell.* Depending on which newer edition of each book is the more recent, the other could be out of date on certain points.

deDuve, Christian. *Blueprint for a Cell: The Nature and Origin of Life.* Burlington, N.C.: Neil Patterson, 1991.

A Nobel laureate, who won for discoveries about the functional and structural organization of the cell, explains how cells function and how life might have begun.

Grobstein, Clifford. *Science and the Unborn.* New York: Basic Books, 1988.

The author, a specialist in biology and public policy, discusses the moral and ethical status of the human embryo and fetus in light of modern biology's understanding of the developmental process.

Kessel, Richard G., and Randy H. Kardon. *Tissues and Organs: A Text-Atlas of Scanning Electron Microscopy.* San Francisco: Freeman, 1979.

This is an album of photographs of actual cells and tissues, showing what they look like in the body. The pictures were taken with a scanning electron microscope, which yields superb images of the architectural features of cells.

Moore, Keith L. *The Developing Human.* 4th ed. Philadelphia: Saunders, 1988.

One of the best books on fertilization and embryonic development of the human body, this also covers common birth defects.

Sadler, T. W. *Langman's Medical Embryology.* 5th ed. Baltimore: Williams & Wilkins, 1985.

This is an excellent text on human fertilization and development, complemented by colorful diagrams.

Vander, Arthur J., et al. *Human Physiology: The Mechanisms of Body Function.* 6th ed. New York: McGraw-Hill, 1995.

An outstanding book on human physiology, whose first few chapters cover cell and molecular biology in moderate detail, with the rest of the book explaining all the major organ systems.


At these sites on the World Wide Web, browsers can learn a variety of things, ranging from the basics to the latest increments in research findings. As of early 1996, these were some of the most interesting sites to be found, and each offers links to other sites.

*Cell & Molecular Biology On-line.* www.tiac.net/users/pmgannon/cool.html

This is a central listing of links to many relevant resources on the Web.

*Cells Alive!* www.comet.chv.va.us/quill/

This gallery of pictures of various types of cells includes movies of cells crawling, macrophages eating bacteria, and the like.

*The Dictionary of Cell Biology.* www.mblab.gla.ac.uk/~julian/Dict.html

Searchable.

*Genetics and Cell Biology Courses On-line.* lenti.med.umn.edu/~mwd/courses.html#GENETICS

Several institutions have put courses of instruction in cell and molecular biology on-line.

*Primer on Molecular Biology.* www.gdb.org/Dan/DOE/intro.html

This on-line course begins with the basics of DNA structure and moves into a discussion of the progress and social implications of the Human Genome Project.

# INDEX ●

15–21; cells express specializations, 165
cell cycle: general process, 138–
143; in cancer, 236, 238
cell division, 117–143; as fountain
of youth, 247
cell membrane, 63
cell motility, 44–61
cell theory, 6–9
cell types, 11
centriole: structure and hypothetical cell "brain," 58, 59; role in
cell division, 131, 133
centrosome: structure, 57; in cell
division, 131
centrosphere, 57, 58
*Chance and Necessity,* 24
channel, portals through cell
membrane, 63, 64, 66
chaperone molecule, 105
chicken embryo, 45
cholera, 67
cholesterol: essential to membranes, 65; how transported
into cell, 69, 93
chromatin, 85
chromosome, 83, 84; structure, 85,
124, 136; number in egg, 146;
broken, 235
cilia, 42; transport ovum, 146; in
windpipe, 214
clathrin, 69–70
Cleaver, James, 232
Clegg, James S., 253–254
codon: defined, 91; ribosome
reads it, 102–103; reading
frame of, 126
cold, common, 5
collagen, 16, 64; in muscle, 194;
in wound healing, 207
compaction, 153
complement, 216
complementarity: principle explained, 82; diagrammed for

DNA, 89, 90; explained for
RNA, 98; in DNA replication,
126
conception, a misnomer, 144
"conception, life begins at," misleading phrase, 144
conceptus, 144, 151
consciousness, 59
contraception, new frontiers in research, 144, 148
coronary thrombosis, 208
corpus luteum, 159; secretes hormones in pregnancy, 186
Crick, Francis: discoverer of double helix, 22; speculates about
consciousness, 59, 61; Nobel
Prize, 88
Crowe, John H., 254
cryptobiosis, 249–255
cumulus cells, 146, 148
curare, 199
cyclin, 141–142
cystic fibrosis, 82
cytomegalovirus, 213
cytoplasm, 63
cytosine, 89
cytoskeleton, defined, 48; role in
cell motility, 53, 57
cytotoxic T cell, 220

Dartmouth College, 36
Darwin, Charles, 88
death, 247–266
desmosome, 153; diagrammed,
155; in wound healing, 209
development, human: fertilization, 144–160; growth of true
embryo, 161–188
diabetes, type I, as autoimmune
disease, 227
*Dictyostelium,* 8
differentiation, starts in development, 154–156
DMSO, 17

filopodium, 51–53
flagellum, 39, 42; of sperm, 147
Flemming, Walther, 9
fluid mosaic, 68
Folkman, Judah, 243
free radicals: as cause of aging, 262; antioxidants, 262
freezing cells, 3, 19
Fuller, Buckminster, 69

Galvani, Luigi, 196
ganglia, 46
gap junction, 154 (illus. 155)
Gardner, Charles A., 172
Gatorade, 200–201
Gaucher's disease, 82
Gelboin, Harry, 231
genes: how they work, 81–116; number of, 84; regulation of, 92–96; transcription, 97 99; translation, 99–103; first expression after fertilization, 152–153; not whole story in development, 172–173; mutations affecting development, 173; reading frame, 202; for antibodies, 222–223
genetic code, how it works, 88–91
Geron Corporation, 241
giant axon, squid, 36
Gilbert and Sullivan, 33
Gilbert, Walter, 107
glucose, 74–76
glycerin, 17
God, 6
Goldman, Robert D., 27, 92
Goldstein, Joseph, 68–71
Golgi apparatus, 111–114; lacking in sperm, 146
Golgi, Camillo, 111
Gorbsky, Gary, 134
Gottleib, Max, 10
guanine, 89

Harley, Calvin B., 241
Harrison, Ross G., 15
Harvard University, 107, 140, 243
hay fever, 5
Hay, Rob, 3, 21
Hayflick, Leonard, 256–257
heart attack, 208
heart disease, 5
HeLa cells, 4, 16
helper T cell, function, 218–221
heme, 110
hemoglobin: defective in sickle cell anemia, 81; defect in gene first to be found, 87; structure and synthesis, 100, 108, 110; fetal hemoglobin in sickle cell anemia 116; gene in embryonic development, 186; in muscle, 196
hepatitis B, 240
Hippocrates, 242
histamine: in inflammation, 214; in allergy, 227
HIV (human immunodeficiency virus), 220–221
homeobox gene, 175–176
Hooke, Robert, 6, 7
hormone, effects spread through gap junctions, 154
human leukocyte associated antigen, 225
Hunt, Tim, 139–141
Huntington's disease, gene found, 82; 182
Huxley, Thomas Henry, 33
hydrogen peroxide, 216
hypothalamus, 187

immune system, 107; 212–228
Imperial Cancer Research Fund, 141
in vitro fertilization, 151–152
infectious diseases: described, 212, immune system response to, 212–228

inflammation, 214–217

inner cell mass: destined to be true embryo, 156; first steps in development, 165

Inoue, Shinya, 118

insulin, 5, 64

integrin, 54

interleukin-2, 221

intermediate filament: type of cytoskeleton, 48; as possible signaling system, 92

intervening sequence, 98

intron: defined, 99; prevalence in genes, 106

Jacob, Francois, 118

Jenner, Edward, 224–225

Kartagener's syndrome, 147

kidney cell, 11

killer cells, 13

kinesin: discovery, 37–40, 63; possible role in mitosis, 135

kinetochore, 133–136; in fertilization, 151

Kirschner, Marc, 134

Korn, Edward, 48

Krebs cycle, 78

Krebs, Hans, 78

Kunkel, Louis M., 201–203

Lacks, Henrietta, 4, 246

laminin, 54

Lawrence Berkeley Laboratory, 164

LDL: discovery of cholesterol transport mechanism, 69–73; structure of gene, 106–107

LDL receptor gene, 93

Le Douarin, Nicole M., 180

Leeuwenhoek, Antoni van, 251

leukemia, 234

Levi-Montalcini, Rita, 209–210

Lewis, Sinclair, 10

life, is antientropic, 247

*Life* magazine, 161

"life, begins at conception," misleading phrase, 144

lipofuscin, as cause of aging, 259

liver cancer, 239–240

liver cell, 11

*Lives of a Cell, The,* 103

Loeb, Jacques, pioneers mechanist view, 9–11, 24

Lou Gherig's disease (ALS), 33

lovastatin, 65

low-density lipoprotein. *See* LDL

lymphoma, 234

lysosome, 28

macrophages: free-roaming cell type, 44, 45; body's garbage collectors, 60; means of eating, 71–72; in wound healing, 208; in immune system, 213–214, 217; develop from monocyte, 221

major histocompatibility complex, 225–226

mammary cells, in vitro, 164

Mangold, Hilde, 169

Margulis, Lynn, 79

Marine Biological Laboratory: center of early cell biology, 9; scene of kinesin discovery, 36; scene of microscope development, 118; scene of cell cycle discoveries, 139

mast cells: in inflammation, 214; in allergy, 227

Masui, Yoshio, 139

*Mechanistic Conception of Life, The,* 9

mechanistic view, 9

Medical Research Council (of UK), 169

megakaryocyte, 206